Control of Liquid–Liquid Extraction Columns

Topics in Chemical Engineering

A series edited by R. Hughes, University of Salford, UK.

Additional volumes in preparation

ISSN: 0277-5883

CONTROL OF LIQUID–LIQUID EXTRACTION COLUMNS

By Kaddour Najim

Institut National Polytechnique de Toulouse, France

GORDON AND BREACH SCIENCE PUBLISHERS

New York   London   Paris   Montreux   Tokyo   Melbourne

Gordon and Breach Science Publishers

Post Office Box 786
Cooper Station
New York, New York 10276
United States of America

Post Office Box 197
London WC2E 9PX
England

58, rue Lhomond
75005 Paris
France

Post Office Box 161
1820 Montreux 2
Switzerland

3-14-9, Okubo
Shinjuku-ku, Tokyo
Japan

Private Bag 8
Camberwell, Victoria 3124
Australia

Library of Congress Cataloging-in-Publication Data

Najim, K.
    Control of liquid-liquid extraction columns

    (Topics in chemical engineering, ISSN 0277-5883 ;
v. 5)
    Bibliography : p.
    Indcludes index.
    1. Extraction (Chemistry) 2. Chemical process
control. I. Title. II. Series.
TP156.E8N24 1988   660.2'8424   88-32004
ISBN 2-88124-703-2 (France)

Contents

APPENDIX A

APPENDIX B

To Michelle and Pierre Lofti

Introduction to the Series

The subject matter of chemical engineering covers a very wide spectrum of learning and the number of subject areas encompassed in both undergraduate and graduate courses is inevitably increasing each year. This wide variety of subjects makes it difficult to cover the whole subject matter of chemical engineering in a single book. The present series is therefore planned as a number of books covering areas of chemical engineering which, although important, are not treated in any length in graduate and postgraduate standard texts. Additionally the series will incorporate recent research material which has reached the stage where an overall survey is appropriate, and where sufficient information is available to merit publication in book form for the benefit of the profession as a whole.

Inevitably, with a series such as this, constant revision is necessary if the value of the texts for both teaching and research purposes is to be maintained. I should be grateful to individuals for criticisms and for suggestions for future editions.

R. HUGHES

Preface

There has been a growing awareness over the last decade that adaptive control can produce improved performance in real applications, mainly as a result of considerable advances in the understanding of adaptive control theory as well as in computer technology. Many feasibility studies have been performed and several adaptive controllers are available on the market.

Faced with new requirements, it seems to have become necessary to put at the reader's disposal a book dealing with basic and up-to-date aspects related to identification theory, adaptive control and learning control, as well as with aspects related to the implementation of these different techniques for process control, such as liquid–liquid extraction columns, which are representative of the difficulties and problems to be met in industrial control design applications. On the other hand, the use of liquid–liquid extraction for industrial process is gaining increasing attention in several industries such as the petrochemical, nuclear, hydrometallurgical and food industries, and is also gaining interest in the pharmaceutical industry field and in fine chemistry.

The identification deals with estimation of model parameters based on observed data from the systems. These models can be derived from:

— physical, chemical and biological considerations where the parameters to be identified generally have physical interpretations (heat transfer coefficients, etc.)

— standard input-output representations where the parameters to be estimated are such that the model relates perfectly to (in some sense, e.g., least squares, etc.) the observed data.

These techniques are special tools not only for liquid–liquid extraction columns model parameters identification. The formulation and the solutions to the identification problem developed in this book are general and can be used without restriction for any processes.

In the chapter concerning adaptive control, we present algorithms which can be used to control any system with complex dynamics. We have voluntarily omitted treatment of the model reference approach because this technique is only applicable to minimum phase systems. For discrete systems, the minimum phase property is in practice, more the rule than the exception, independent of whether the continuous process transfer function has unstable zeros or not. The robust adaptive control strategies: pole placement, linear quadratic Gaussian controller and generalized predictive control, are presented.

A very promising control approach, which is based on learning systems, is also detailed. In this control approach all the knowledge (know-how, correlations, models, etc.) concerning the process to be controlled can easily be taken into consideration. This control approach is connected intimately to artificial intelligence which is a field of increasing development.

Finally, this book contains several Fortran subroutines to help the reader in implementing identification and control strategies without difficulty.

K. NAJIM

IDENTIFICATION

1. INTRODUCTION

Two methods are commonly used to represent systems knowledge: the abacus or diagram and mathematical models. This systems knowledge plays an important role in a variety of fields, e.g., for plant design, sizing and costing calculations, control purposes, optimal network synthesis configuration, forecasting, operator training, and also for quantitative analysis of phenomena in research areas.

The models are mathematical representations of process behaviour. They represent the interactions existing between the different variables (temperature, flow rate, etc.) acting on the process. For example, the dryer process among others leads to a relation between energy and the moisture content of the product to be dried.

The procedures for model elaboration can be classified into the following three categories:

1. Physical model

2. Input-output representation

3. Learning automata

In the first category, the models are carried out from fundamental physical and chemical (and biological, etc.) laws, such as the laws of conservation of mass, energy, and momentum. Because of the complexity of the process or the phenomena to be mathematically described, the engineer makes some assumptions to simplify the description. In spite of these hypotheses, the resulting models are often described by non linear and coupled partial differential equations (Najim et al., 1976a, 1976b).

In the second category, the model (transfer function, state representation, etc.) which is developed by control engineers who pay little attention to the process to be controlled. The cause-and-effect relationship between input and output signals consists for continuous systems of a set of algebraic, differential equations, partial or not, which can be discretized for simulation or control strategy implementation on a digital computer. There also exist systems which are fundamentally discrete, such as the problem related to the optimization of the number of stages in a rocket or to the increase of production on a breeding farm.

In the third category, the process and the controller are modeled by the same system: a learning automaton operating in a random environment. The environment corresponds to the plant to be controlled. The learning automaton and the medium (environment) are connected in a feed-back loop where the input to one is the output to the other. The learning automaton consists of a performance evaluation unit, and a stochastic automaton with varying structure (El-Fattah and Foulard, 1978; Lakshmivarahan, 1981). A probability distribution vector is associated to the control variable which is discretized in a set of values. Using a reinforcement scheme and based on the performance unit evaluation and environment responses, the automaton generates an action which tends, with time, to optimal control action, in the sense of minimizing a predefined cost function.

In this chapter we shall be concerned with the problem of parameter estimation. In fact, the models developed for processes depend on parameters for which numerical values must be assigned so that the models can simulate in some way the process behaviour, i.e., the process mathematical representation must be able to predict correctly (in some sense) the process output. Sometimes, these parameters have physical significance (heat exchange coefficients, residence time, kinetic coefficients, etc.) and may be of great importance for process design or for improving the performance of existing processes.

Identification deals with the parameter estimation problem based on the information collected on the process, i.e., the input-output measurements. These parameters cannot be directly obtained from input-output measurements. In the signal processing field, engineers are also concerned with signal identification and pattern recogition. The optimization of an industrial chemical system, which is a set of functional elements (an apparatus) can be stated as an identification problem to which the resolution consists of the search for a collection of parameter of the system, characterizing its operation and providing the extreme value of the optimization criterion.

2. PHYSICAL MODEL

Based on physical and chemical (or biological) laws, mathematical models can be developed from energy, mass and momentum balances, i.e., the general principle of conservation; for example, for chemical processes which generally involve heat and mass transfers phenomenon. This means that if we consider a closed volume, the accumulation is equal to the inflow minus the outflow. We shall present in the following an example of the elaboration of a model. The process which will be considered is a phosphate drying furnace. Drying is an irreversible thermal phenomenon associated with large energy losses, which constitutes an important source of energy consumption. Indeed, the drying operation represents more than the quarter of the cost price of the raw material. If the drying operation is carried out economically, this can save a considerable amount of energy. It is essential to have a dryer of appropriate dimensions and to fit it with high-performances controllers.

The phosphate drying furnace depicted in figure 1 is mainly composed of the following elements: a combustion unit, a rotary drying tube and a dust chamber.

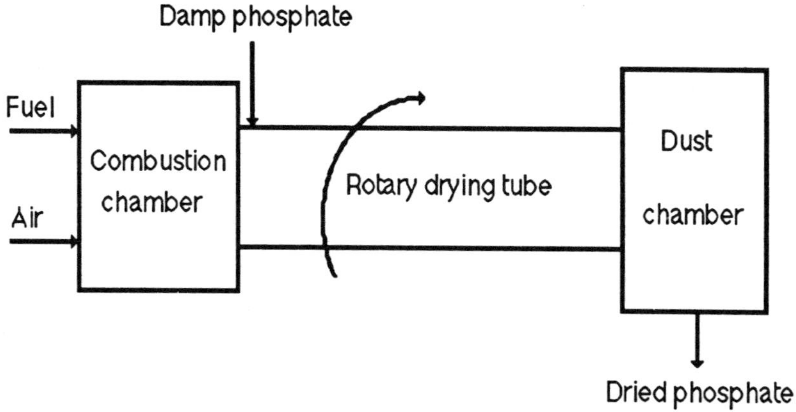

Figure 1. Schematic diagram of the furnace

In the combustion unit the fuel, initially preheated to about 100 °C and pulverized by means of stream (to facilitate its combustion), is burned. The oxygen necessary for combustion is provided by the primary air input. The heat produced by the combustion is taken out by the secondary air which is induced by a ventilator.

The hot combustion gases, at a temperature of about 800 °C, pass to the drying tube, where they are mixed with the damp phosphate which enters at the front of the rotary tube. The moisture content of the damp phosphate is about 15%. The thermal exchange between the product and the hot gases, and the forward movement of both the product and the gases, are facilitated by the constant rotation of the horizontal tube and the helicoidal cascades on its inner surface, and also by the lower air pressure caused by the draught ventilator at the end of the plant. The phosphate and hot gases flow in the same direction, unlike the situation in a cement furnace. The dust chamber is mainly made up of shelved tubes whose function is slow down the fine phosphate particles which are collected and carried to join the final product (moisture content of about 1%) received by the main conveyer.

The main role of the ventilator is to create a reduction in pressure at the head of the drying tube in order to induce a secondary air flow (to carry the calories) and to prevent the phosphate from being trapped in the drying tube. The chimney action evacuates the hot gas from the furnace.

In practice the process is influenced by a number of random perturbations, such as fluctuations in the input flow and the moisture content of the wet material and other varying parameters (ambient temperature, fuel properties, etc.). Modern control methods require a good mathematical description of the process in question. The process structure may be derived from basic physics, whereby the parameters can be ascribed physical significance. Taking this to the extreme, the model may simply be chosen to give the "best" relationship between input and output measurements.

To model this furnace we assume that all the variables

considered in the following are constant in z and y directions and that they depend only on the space variable x and on the time variable t. We shall also assume that the specific heats and the velocities of the gases and the phosphate, as well as the sheet transfer coefficients, are all constant. To calculate the partial gas flows, the composition of the fuel and the gases may be supposed to be constant. The diffusion of the water through the phosphate and through the gas, which can be described by a partial differential equation of second order, may be approximated by a term function of the vaporization rate and of the latent heat.

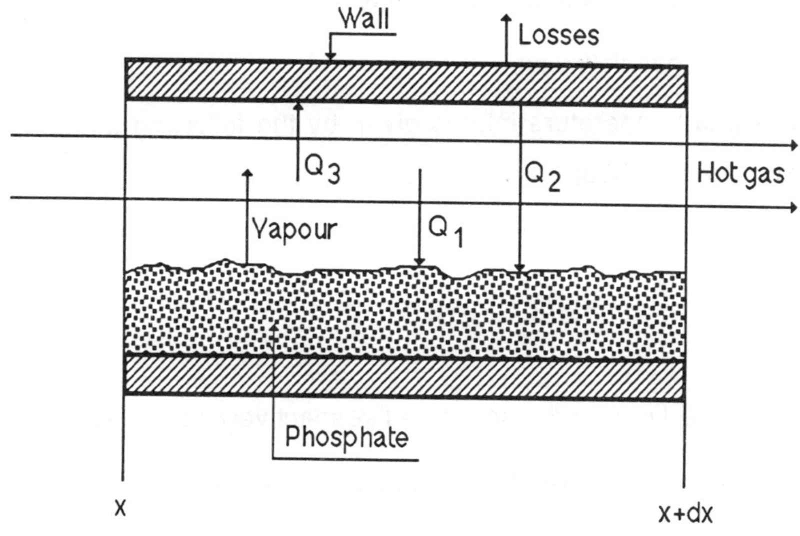

Figure 2. Heat transfer in the furnace

Based on heat and mass transfer (figure 2), a mathematical

model was developed by Najim et al. (1976):

$$C_p \partial[D_p T_p]/\partial x + C_s(1/V_p)\, \partial[D_p T_p]/\partial t = Q_1 + Q_2 - R_{H2O}\lambda \qquad (1)$$

$$C_g \partial[D_g T_g]/\partial x + C_g(1/V_g)\, \partial[D_g T_g]/\partial t = -Q_1 - Q_3 + R_{H2O}\lambda \qquad (2)$$

$$\partial[D_p]/\partial x + (1/V_p)\, \partial[D_p]/\partial t = -R_{H2O}\lambda \qquad (3)$$

$$\partial[D_g]/\partial x + (1/V_g)\, \partial[D_g]/\partial t = +R_{H2O}\lambda \qquad (4)$$

The boundary conditions are:

$T_p(0,t) = Ta$ (ambient temperature)

$T_g(0,t) = Tg0$ (gas temperature inlet)

$D_p(0,t) =$ phosphate flow at the inlet

$D_g(0,t) =$ gas flow inlet

The gas temperature inlet is given by the following enthalpy equation (Perry, 1963):

$$Q = \int_{Ta}^{Tg0} [C_1 D_1 + C_2 D_2 + C_3 D_3 + C^*_1 D^*_1 + C^*_2 D^*_2]\, dT \qquad (5)$$

where C_i and D_i, $i = 1,3$ correspond respectively to the specific heats and the partial flow of the products of fuel combustion (CO_2, H_2O and SO_2), and C^*_i, $i = 1, 2$ correspond respectively to the specific heats and the partial flow of O_2 and N_2.

C_g, C_p, D_g, D_p, T_g, T_p, V_g and V_p are the specific heat, the flow rates, the temperatures and the velocities of the gas and phosphate respectively. R_{H2O} and λ are the vaporization rate and the latent heat of vaporization respectively. Q_1 and Q_2 are the heat quantities transferred to the phosphate respectively by the gas and by the inner wall of the drying tube. Q_3 is the heat quantity transferred to the inner wall of the furnace by the gas.

Equations (1) and (2) express the energy conservation, while equations (3) and (4) express the mass balance. These two last equations show that the vapour lost by the damp material supplies the gas.

The different heat quantities transferred are functions of the temperature gradients and of the furnace parameters such as the mean depth of solids inside the furnace, heat transfer coefficient between gas and wall, etc.

These equations associated to boundary conditions contitute the drying furnace model. They are complex partial nonlinear differential equations and can not be used for control purposes. On the other hand they can be used for dryer sizing. They show that the fuel flow and the moisture content of the dried phosphate are the most important representative variables for behaviour of the furnace. They can also be used as good representations of the process dynamics, to try out the performances of control strategies derived from input output models (black box models).

3. LINEAR DISCRETE MODEL

There is increasing interest in the issue of digital control implementation owing to the rapid development of high speed digital computers. The dynamics of processes may be represented by several types of models (differential equations, partial differential equations, etc.). For a digital controller it is convenient to use discrete models. Several techniques are commonly used to transform the continuous model to the time discrete form with sampling time T. As an example, let us derive a discrete version form of a common regulator, the P.I.D. .

The transfer function of the continuous P.I.D. is:

$$F(s) = k\{1 + T_i s + T_d s/(1+ \alpha s)\} \tag{6}$$

where s represents the Laplace operator.

The regulator parameters are: k, kT_i and kT_d which are respectively the proportional integral and derivative effects. The parameter α is usually fixed by the manufacturer and can not be changed by the user.

The discrete transfer function of the P.I.D. depends on the approximation rules used to represent in a numerical manner the integral and the derivative parts of the function involved in the P.I.D.

A common expression used in the literature for a discrete P.I.D. regulator is:

$$H(q^{-1}) = (a_0 + a_1 q^{-1})/(1 - q^{-1}) + b_0(1 - q^{-1})/(1 + b_1 q^{-1}) \qquad (7)$$

where q^{-1} is the backward shift operator ($q^{-j}y(t) = y(t - jT)$) and T represents the sampling period.

This expression uses the general solution of differential equations with second members for the derivative part of the P.I.D. supposed to be constant between two consecutive times, and a simple numerical approximation for the integral term. Other expressions for discrete P.I.D. regulators can be obtained using other approximation techniques for continuous systems.

In Table 1, the relations existing between the discrete and the continuous P.I.D. regulator parameters are given (Wittenmark, 1979).

Table 1. Discrete and continuous P.I.D. regulator parameters

$a_0 = k$	$k = a_0$
$a_1 = k(T/T_i - 1)$	$T_i = a_0 T/(a_0 + a_1)$
$b_0 = kT_d(1 - \exp(T/\alpha T_d))/T$	$T_d = b_0 T/(k(1 + b_1))$
$b_1 = -\exp(-T/\partial T_d)$	$\alpha = -k(1 + b_1)/(b_1 \ln(-b_1))$

4. THE   LEAST SQUARES METHOD

Consider a linear dynamical system with control signal u(t) and output signal y(t) governed by the following differential equation:

$$y(t) + \sum_{i=1}^{n} \alpha_i y^i(t) = \sum_{i=0}^{m} \beta_i u^i(t) + p(t) \tag{8}$$

where p(t) is the disturbance (measured or not) acting on the considered process, though it would also include the effect of modelling errors. The discretization of equation (8) with a sampling period T leads to the linear discrete equation

$$y(t) + \sum_{i=1}^{n} a_i y(t-iT) + = \sum_{i=0}^{m} b_i u(t-kT-iT) + p(t) \tag{9}$$

In equation (9) we have introduced explicitly the dead-time (transport delay) kT (supposed to be an integer multiple of the sampling time T).

Let us introduce the backward shift (or delay) operator q^{-1} .

$$q^{-j}y(t) = y(t - jT) \tag{10}$$

Then equation (8) can be rewritten as

$$A(q^{-1})y(t) = q^{-k}B(q^{-1})u(t) + p(t) \tag{11}$$

where $A(q^{-1})$ and $B(q^{-1})$ are polynomials in the delay operator:

$$A(q^{-1}) = 1 + a_1 q^{-1} + \ldots + a_n q^{-n}$$

$$B(q^{-1}) = b_0 + b_1 q^{-1} + \ldots + b_m q^{-m} \tag{12}$$

Let us define the parameters and the observation (input output data) vectors

$$\theta = [\, a_1, a_2, \ldots, a_n; b_0, b_1, \ldots, b_m \,]^T$$

$$X(t) = [-\, y(t - T), \ldots, -\, y(t - nT); u(t - kT), \ldots, u(t - kT - mT)]^T \tag{13}$$

Then equation (6) can be rewritten as

$$y(t) = \theta^T X(t) + p(t) \tag{14}$$

This relationship is commonly used for identification and control purposes. It represents a class of processes. To characterize in some way a specific process, numerical values must be assigned to the model's parameters to obtain a similitude of the properties of the model and the properties of the considered system.

Based on the available data (input-output measurements), our aim is to obtain a good estimate (in the least squares sense) of the vector parameter θ.

The least squares estimation minimizes the following quadratic criterion:

$$J = \sum_{i=0}^{t} [y(i) - \theta^T X(i)]^2 \tag{15}$$

We will denote the optimal value θ, which leads to the extreme of function (15) , by θ^*_t. Note that the extremum of interest to us is a minimum. Finding θ^*_t reduces to solution of the equation that is obtained by setting the gradient of J equal to zero, i.e.

$$\partial J/\partial\theta = 0 \tag{16}$$

Combining (15) and (16) yields

$$\theta^*_t = \left[\sum_{i=0}^{t} X(i)XT(i)\right]^{-1}\sum_{i=0}^{t} y(i)X(i) \tag{17}$$

The second derivative matrix is:

$$\partial^2 J/\partial\theta^2 = \sum_{i=0}^{t} X(i)X^T(i) \tag{18}$$

showing that this extremum (17) corresponds to the global minimum of the sum of the squares of the deviations between the measured output and the one predicted by the model.

$$\text{Let}\quad P(t) = \sum_{i=0}^{t} X(i)X^T(i) \tag{19}$$

Equation (17) becomes

$$\theta^*_t = P^{-1}(t)\sum_{i=0}^{t} y(i)X(i) \tag{20}$$

In order to illustrate the formulation of the identification problem just described, let us consider two examples.

Many correlations have been proposed in the literature for physical properties.

For simulation dynamic processes, simplified relations based on polynomial regression whose parameters can be identified by the least squares method are usually used to reduce the computational time. Several relations are given in tables or charts which are useful to rapidly derive approximate results which give the primary information about the problem under consideration, which may serve engineers in their investigations. Good results are generally obtained with second degree polynomials. Therefore, the number of parameters to be identified does not exceed 3.

Example 1:

The vapour saturation pressure P_s (mmHg) may be estimated by a second degree polynomial in the temperature range: $10°C \leq T \leq 40°C$. The correspondence between these two physical parameters is given in the form of a table of numerical values (Table 2).

$$P_s = a_0 + a_1 T + a_2 T^2$$

Table 2. Numerical values of the couple vapour saturation
pressure and temperature

Temperature C°	10	12	14	16	18	20	22	24	26
Vapour saturation pressure	9.20	10.51	11.98	13.63	15.47	17.53	19.82	22.37	25.20

Temperature C°	28	30	32	34	36	38	40
Vapour saturation pressure	28.34	31.82	35.66	39.90	44.57	49.70	55.33

The least squares method gives : $\hat{a}_0 = -114.76$; $\hat{a}_1 = 8.54$; $\hat{a}_2 = -0.032$

Example 2:

When digital computers are used in a feed-back loop for
control purposes, the characteristics of the sensors (thermocouple,

flow meter, etc.) and of the actuators (valve, motor, etc.) which are generally nonlinear, and given by table or diagram, are modeled by polynomials for which the presented method can be used for parameter estimation.

Remarks:

1. Owing to the fact that the measurements are disturbed, the vector parameter estimation θ^{*}_{t} is random. This estimation can be an unbiased estimate of the true vector parameters θ. An estimator is said to be unbiased if the mathematical expectation of the parameter estimation is equal to the true parameter. The least squares estimate of θ is unbiased if the perturbation p(t) has zero mean and if p(t) and the data are statistically independent.

Notice that the statistical independence of the observations and the zero mean p(t) is sufficient but not necessary for carrying out unbiased estimation of the vector parameters.

2. The least squares estimator (17) requires a matrix inversion which is undesirable. When an additional measurement became available, or when the parameter number increases, the computational time and the necessary memory required for calculation will be very great. Thus, if the inversion can be circumvented, it is advisable to do so.

In the following, we shall present an identification scheme for estimation of parameters, when the the perturbation $p(t)$ is a linear function of a sequence of independent zero mean random variables with finite variance (normally distributed random variables), and develop the recursive form of the least squares estimator algorithm.

5. THE GENERALIZED LEAST SQUARES METHOD

Many different equations can be used to describe the dynamic relationship of a given process. Let us consider the following representation:

$$A(q^{-1})y(t) = q^{-k}B(q^{-1})u(t) + C(q^{-1})e(t) \tag{21}$$

where $C(q^{-1})e(t)$ models the perturbation acting on the process (noise measurements, etc.), $\{e(t)\}$ is a sequence of zero mean random variables with a finite variance. The polynomial $C(q^{-1})$ is given by:

$$C(q^{-1}) = 1 + c_1 q^{-1} + \ldots + c_n q^{-n} \tag{22}$$

The considered system (equation (21)) is illustrated by the block diagram in figure 3.

Equation (21) may be written in the following form, as indicated in figure 4:

$$A(q^{-1})y^f(t) = q^{-k}B(q^{-1})u^f(t) + e(t) \tag{23}$$

where $y^f(t)$ and $u^f(t)$ are the filtered inputs outputs given by:

$$y^f(t) = [1/C(q^{-1})]y(t), \quad u^f(t) = [1/C(q^{-1})]u(t) \tag{24}$$

where

$[1/C(q^{-1})]$ is the transfer function of the filter.

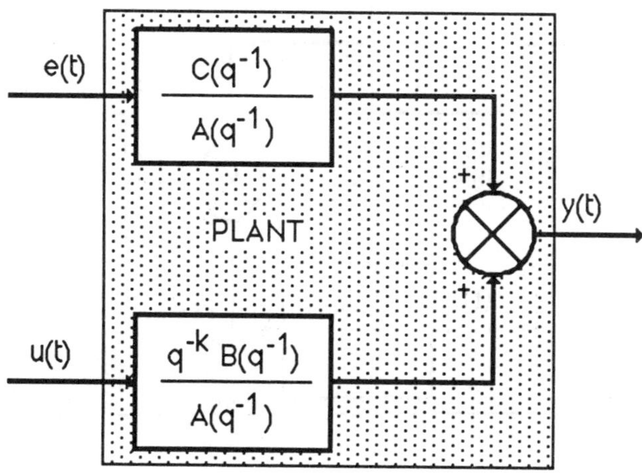

Figure 3. Process and disturbance schematic diagram

If the filter is known, the least squares method can be used to estimate the model's parameters. Generally, this filter is unknown, so it is also necessary to estimate its design parameters (the coefficients c_i, $i = 1, n$). In the following we detail the different steps to be performed to simultaneously estimate the parameters of the polynomials $A(q^{-1})$ and $B(q^{-1})$ and the coefficients of the filter $[1/C(q^{-1})]$.

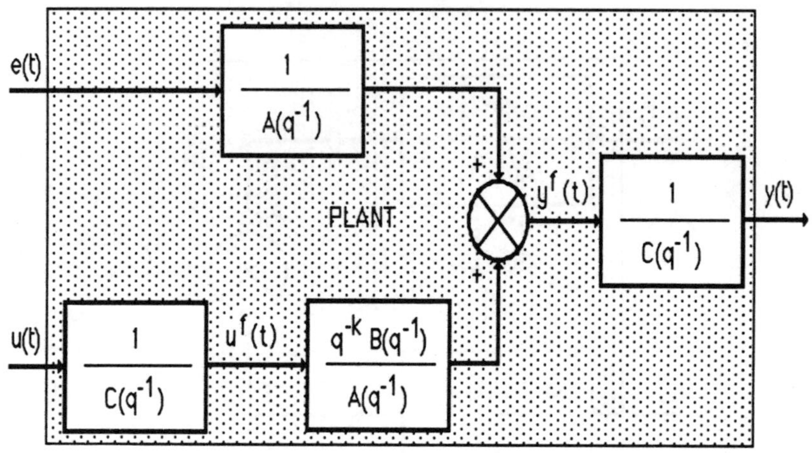

Figure 4. Filtered input-output system

Step 1 : Select an a priori value of the coefficients c_i

Step 2 : Compute the filtered values of the input u(t) and of the output y(t) (as indicated by equation (24))

Step 3 : Estimate the parameters a_i and b_i (the polynomials $A(q^{-1})$ and $B(q^{-1})$ coefficients) using the least squares method

Step 4 : Evaluate the residual error $\epsilon(t)$ (the prediction error)

$$\epsilon(t) = A^*(q^{-1})y(t) - B^*(q^{-1})u(t)$$

The polynomials $A^*(q^{-1})$ and $B^*(q^{-1})$ have the same structure as the polynomials $A(q^{-1})$ and $B(q^{-1})$. The unique difference lies in the fact that their coefficients are replaced by the estimations performed in step 3

Step 5 : Estimate the filter parameters

The residual is also given by:

$$\epsilon(t) = [1/C(q^{-1})] \, e(t)$$

It can be also expressed as follows:

$$\epsilon(t) = -c_1\epsilon(t) - c_2 \, \epsilon(t) - \ldots - c_n \, \epsilon(t) + e(t)$$

$\{e(t)\}$ is a sequence of independent random variables, then the least squares method, if used, leads to unbiased estimates of the filter parameters.

Go to step 2

There exist other estimation algorithm for solving the identification problem formulated above. Among them we can mention the instrumental variable method, the maximum likelihood approach (Mendel, 1973); Ljung and Söderstöm, (1983)).

The following sections deal with recursive identification techniques which have several fixes and are commonly used in the design of adaptive controllers.

6. THE RECURSIVE LEAST SQUARES METHOD

Consider a new input-output measurement and denote by θ^*_{t+1} the corresponding parameter estimator.

$$\theta^*_{t+1} = P^{-1}(t+1) \sum_{i=0}^{t+1} y(i)X(i) \tag{25}$$

We are going to express θ^*_{t+1} as a function of θ^*_t.

From equations (17) and (25) we obtain the following expression for θ^*_{t+1}.

$$\theta^*_{t+1} = P^{-1}(t+1)\{ \sum_{i=0}^{t} y(i)X(i) + y(t+1)X(t+1) \} \tag{26}$$

According to (17) we have

$$\theta^*_{t+1} = P^{-1}(t+1) \{ [\sum_{i=0}^{t} X(i)X^T(i)] \theta^*_t + y(t+1)X(t+1) \} \tag{27}$$

Then one finds, using (19) and by adding and substracting the term $X(t+1)X^T(t+1)$:

$$\theta^*_{t+1} = \theta^*_t + P^{-1}(t+1)X(t+1)[y(t+1) - \theta^{*T}_t X(t+1)] \tag{28}$$

<u>Remark:</u>

The estimate correction $\{\theta^*_{t+1} - \theta^*_t\}$ is proportional to the error prediction which is equal to the difference between the measured value $y(t+1)$ and its prediction.

The recursive formula for the covariance matrix P(t) will be derived in the following by using the inversion lemma, which we now state and prove.

Let A, B, C and D be consistently dimensioned matrices; then

$$[A + BCD]^{-1} = A^{-1} - A^{-1}B [C^{-1} + DA^{-1}B]^{-1}DA^{-1} \tag{29}$$

Proof: Premultiply equation (18) by [A + BCD] from the right, obtaining:

$$I = I + A^{-1}BCD - A^{-1}B [C^{-1}+DA^{-1}B]^{-1}D - A^{-1}B[C^{-1}+ DA^{-1}B]DA^{-1}BCD$$

$$I = I + A^{-1}B[C^{-1} + DA^{-1}B]^{-1}\{ [C^{-1} + DA^{-1}B]CD - D - DA^{-1}BCD \}$$

$$I = I$$

which proves the assertion (29).

Let us proceed to evaluate $P^{-1}(t +1)$ as a function of $P^{-1}(t)$. We find

$$P^{-1}(t +1) = \{ P(t) + X(t +1)X^T(t +1) \}^{-1} \tag{30}$$

Compare equations (29) and (30) and make the following correspondence:

$$A = P(t) ; B = X(t +1) ; C = 1 ; D = X^T(t +1)$$

Then:

$$P^{-1}(t + 1) = P^{-1}(t) -$$

$$- P^{-1}(t)X(t + 1)[1 + X^T(t + 1)P^{-1}(t)X(t + 1)]^{-1}X^T(t + 1)P^{-1}(t) \tag{31}$$

Notice that the term $[1+X^T(t +1)P^{-1}(t)X(t +1)]$ is a scalar.

Thus the recursive least squares method can be summarized as follows:

$$\theta^*_{t+1} = \theta^*_t + F(t +1)X(t +1)e(t +1)$$

$$F(t+1) = F(t) - \frac{F(t)X(t+1)X^T(t+1)F(t)}{1 + X^T(t+1)F(t)X(t+1)} \qquad (32)$$

$$e(t+1) = y(t+1) - \theta^*_t{}^T X(t+1)$$

where e(t) represents the prediction error and F(t) the adaptation gain matrix equal to $P^{-1}(t)$ which play, in turn, an important part in recursive identification algorithms.

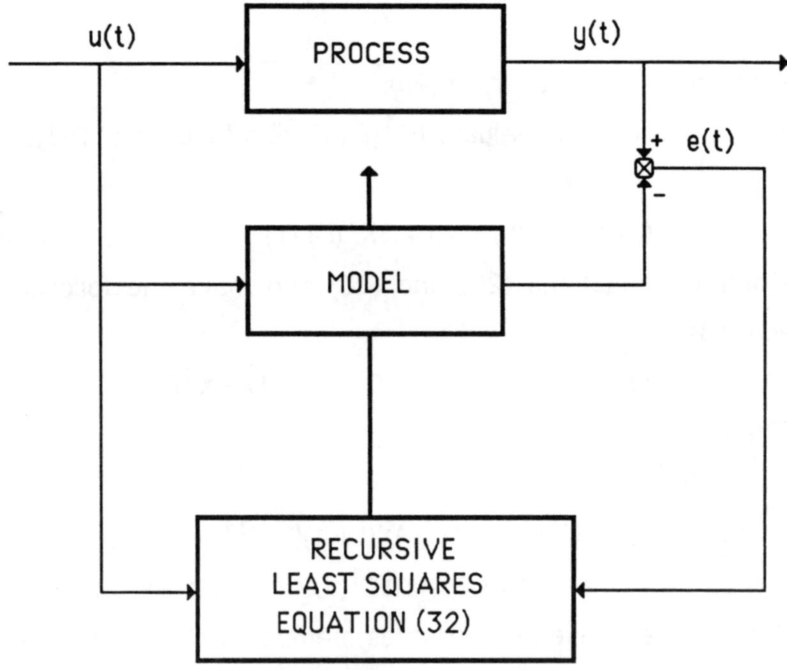

Figure 5. Schematic diagram of the recursive
least squares method

The algorithm defined by equations (32) enables us to recursively calculate the parameter estimate. It does not need any inversion matrix and, consequently, is suitable for real-time identification.

The structure of the obtained identification method is shown in figure 5.

The current parameter estimate is updated using a transformation of the last measurement, multiplied by the gain factor $\{F(.)X(.)\}$.

To be complete, initial conditions related to the vector parameter and to the gain matrix must be associated to the recursion (32).

A priori knowledge of the process can be used to start up the identification procedure. Otherwise, it is common in choosing initial values to take: $F(0) = \beta.I$ and $\theta^*_0 = 0$, where β is a positive constant and I is the unit matrix which ensures positive definiteness of the matrix $F(0)$.

<u>Remark :</u>

The second term on the right-hand side of equation (32) is quadratic. Consequently the gain matrix $F(.)$ decreases monotonically with time, and, when the system is time varying, the prediction error will be great and the correction adaptation term will be very small.

Therefore it appears to be useful to develop an adaptation algorithm with time-varying adaptation gain.

In the following some improvements of the recursive least squares algorithm will be presented in order to avoid this problem.

7. THE WEIGHTED RECURSIVE IDENTIFICATION METHOD

Many reasons can make the dynamics of a process variable :

- The effect of perturbations, like variations in feed flow rate, feed composition, reactant purity, ambient temperature, etc.

- The change of the operating point, for example, the heat transmitted by radiation is negligible at low temperature and increases with rising temperatures. This results in dynamics variation (varying time delay, etc.).

- Owing to wear or aging, the characteristics of the components of plants or the behaviour of some products like the catalysts can also be time varying.

If we analyse criterion (15) we can see that the contribution in this identification index of all the measurements is the same. Nevertheless, for the reasons cited at the beginning of this section, it is necessary to privilege the late information with regard to the old data collected on the process.

To do this, let us introduce the following criterion:

$$J = \sum_{i=0}^{t} \lambda^{t-i} [y(i) - \theta^T X(i)]^2 \qquad (33)$$

Old information is discounted ($t-i \approx t$; $\lambda^{t-i} \approx 0$), in favour of new information ($t-i \approx 0$; $\lambda^{t+N-i} \approx \lambda$) to track time-varying parameters.

The parameter λ represents the forgetting factor (exponential weighting factor). It is usually chosen small and close to one i.e. $0 < \lambda \leq 1$. The measurements received j sampling periods ago has a weight proportional to (Åström, 1983):

$$\lambda^j = e^{jLn(\lambda)} \qquad (34)$$

When λ is close to 1, the time constant for the forgetting factor decay is approximately $\{e^{JLn(\lambda)} \approx e^{J(\lambda+1)}\}$:

$$1/[1-\lambda] \qquad (35)$$

The covariance matrix of the parameter estimate can be written

$$P(t+1) = \sum_{i=0}^{t+1} \lambda^{t+1-i} X(i) X^T(i) \qquad (36)$$

Just as in section III we can of course introduce P(t). Then equation (36) leads to:

$$P(t+1) = \lambda P(t) + X(t+1) X^T(t+1) \qquad (37)$$

For: $A = \lambda P(t)$; $B = D^T = X(t+1)$ and $C = 1$, let lemma (29) apply to (37). We obtain the following alternative expression for P(t +1):

$$P^{-1}(t +1) = \lambda^{-1}\left[\ P^{-1}(t) - \frac{P^{-1}(t)X(t)X^T(t)P^{-1}(t)}{\lambda + X^T(t)P^{-1}(t)X(t)}\ \right] \qquad (38)$$

The foregoing expressions can be summarized by:

$$\theta^*_{t+1} = \theta^*_t + F(t)X(t)e(t)/[\lambda + X^T(t)F(t)X(t)]$$

$$F(t +1) = \lambda^{-1}\left[F(t) - \frac{F(t)X(t)X^T(t)F(t)}{\lambda + X^T(t)F(t)X(t)}\right] \qquad (39)$$

where e(t) represents the prediction error (process output minus the output predicted by the model).

For $\lambda = 1$ we obtain the standard recursive least squares algorithm.

In order to illustrate the usefulness of this forgetting factor, let us consider the following example.

<u>Example 3:</u>

We are interested in identifying parameters in the process model

$$y(t) = a_1 y(t-1) + a_2 u(t-1) + e(t)$$

where y(t), u(t) and e(t) represent respectively the output, the control signal and the error modelisation. The simulations are carried out for: $a_1 = 0.7$; $a_2 = 0.3$. The obtained results are shown in figure 6 for different values of the forgetting factor λ. At time $t = 30$, the parameter a_1 changes from 0.7 to 0.5 .

The evolution of the trace of the gain matrix is depicted in figure 7.

Identification with exponential forgetting of past measurements will continually forget old information whether or not new dynamic information about the plant is available (poor excitation).

When the process to be identified has steady-state or deterministic behaviour, the norm $\{F(.)X(.)\}$ of the adaptation gain becomes very small. The covariance matrix equation (39) is hence given by:

$$F(t+1) = F(t)/\lambda \quad ; 0 < \lambda < 1 \tag{40}$$

That is, the adaptation gain $F(.)X(.)$ grows exponentially and will be very large. Such a problem is referred to as blow-up problem (Åström, 1983).

In order to overcome these problems, it is necessary to use time-variable forgetting factors. For instance, the forgetting factor should go to unity whenever $F(.)X(.)$ is small.

Several suggestions of how to solve this problem have been made. Fortescue et al. (1981) and Wellstead and Sanoff (1981) presented a method using the prediction error as a parameter to adapt the forgetting factor. Ydstie and Sargent (1982) have proposed an algorithm that keeps the information content of the

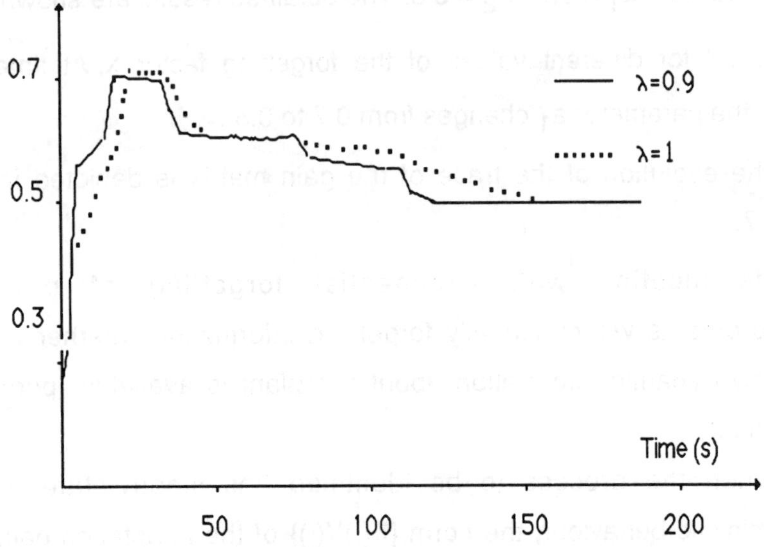

Figure 6. Parameter estimate a^*_1 is obtained in simulation for

different values of the forgetting factor and with $\theta^*_0 = 0.01$,

$F(0) = 500I$.

estimator constant. In Hägglund (1983), an algorithm which enables one to distinguish between disturbances and the others, due to change parameters, is presented. Anderson (1983) has developed an adaptive forgetting factor in recursive identification through multiple models. This approach consists of multiple recursive least squares algorithms running in parallel, each with a corresponding weighting factor. In Favier and Di Martino (1984), the forgetting factor is changed according to measurement of the operating point as in a gain-scheduling approach.

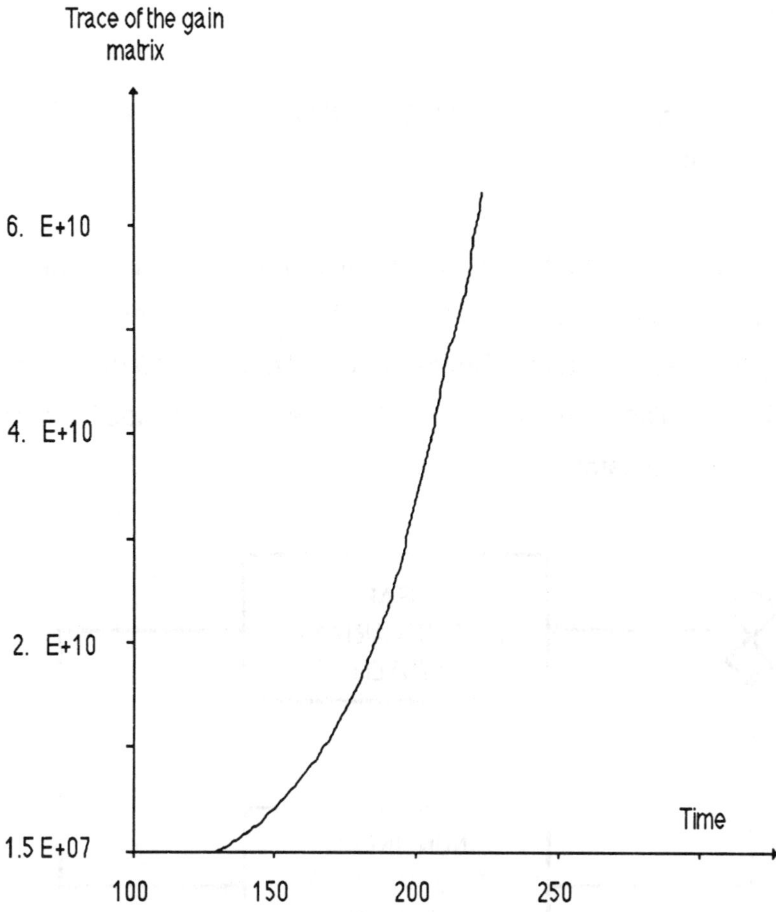

Figure 7. Evolution of the trace of the gain matrix

A general structure of an estimation algorithm using the forgetting concept has been developed by Landau and Lozano (1981). Such a structure easily leads to the well-known constant trace estimation algorithm (Irving, 1979) which will be presented below.

8. LEAST SQUARES METHOD WITH TWO FORGETTING FACTORS

We next consider an identification procedure developed by Landau (Least Squares Method with two Forgetting Factors), using the passivity tools (Desoer and Vidyasagar, 1975; Landau, 1979). This algorithm ensures the global asymptotic stability of the parameter estimator.

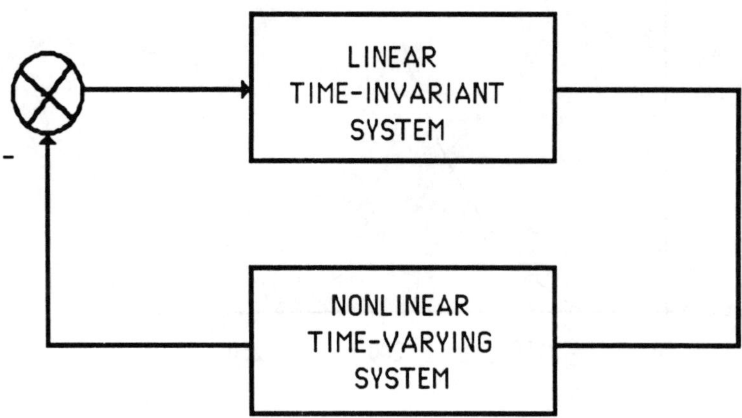

Figure 8. Equivalent non-linear feedback system.

The algorithm analysis was carried out using hyperstability and positivity approaches by transforming the adaptation mechanism into an equivalent non-linear feedback system (figure 8).

Let us introduce in the expression of the adaptation gain matrix (equation (27)) a second ponderation factor ($\lambda_2(t)$). Then it follows that:

$$P(t+1) = \lambda_1(t)P(t) + \lambda_2(t)X(t+1)X^T(t+1) \tag{41}$$

Applying the inversion lemma as for the scheme described in the previous sections for:

$$A = \lambda_1(t)P(t) \; ; \; B = D^T = X(t) \; \text{ and } \; C = \lambda_2(t) \tag{42}$$

one obtains the following adaptation procedure for the gain matrix adaptation gain:

$$P^{-1}(t+1) = [P^{-1}(t) - \frac{P^{-1}(t)X(t+1)X^T(t+1)P^{-1}(t)}{\{\lambda_1(t)/\lambda_2(t)\}+X^T(t+1)P^{-1}(t)X(t+1)}]/\lambda_1(t) \tag{43}$$

where

P(0) is an arbitrary positive definite matrix, $0 < \lambda_1(t) \leq 1$; and

$0 \leq \lambda_2(t) < 2$

<u>Remark :</u>

For different values of the parameters $\lambda_1(.)$ and $\lambda_2(.)$, we obtain various identification algorithms.

If $\lambda_1(.) = \lambda_2(.) = 1$, then equation (43) leads to the classic decreasing gain least squares method. The weighted recursive least squares algorithm corresponds to $\lambda_1(.) = \lambda$ and $\lambda_2(.) = 1$.

Finally, for $\lambda_1(.) = 1$ and $\lambda_2(.) = 0$ we obtain the constant gain adaptation algorithm $\{(P(t)=P(0)\}$ referred to as the projection algorithm (Goodwin and Sin, 1984).

To make sure that the identification algorithm alertly keeps track of changing dynamics and to avoid the problem related to the exponential growth of the covariance-matrix, Irving (1979) has proposed the use of a variable forgetting factor that maintains the trace* of the gain matrix constant. One suitable way to do this as follows:

Let $$\lambda_1(t)/\lambda_2(t) = c \tag{44}$$

where c is a constant value (c = 1 for example). To maintain the trace of the matrix F(.) constant, it follows from the equation (43) that the parameter $\lambda_1(.)$ must be chosen equal to:

$$\lambda_1(t) = \left[\operatorname{tr} F(t) - \frac{\operatorname{tr}\{F(t)X(t+1)X^T(t+1)F(t)\}}{c + X^T(t+1)F(t)X(t+1)} \right] / \operatorname{tr} F(t) \tag{45}$$

where $\operatorname{tr} F(t)$ represents the trace of the matrix F(t).

---

* Let $A = (a_{ij})$ be a n.n matrix. Then $\operatorname{tr} A = a_{11} + a_{22} + \ldots + a_{nn}$

There are other possibilities to obtain an estimator with nondecreasing adaptation gain. For instance we can add a matrix to the covariance one, say

$$F(t+1) = F(t) - [F(t)X(t+1)X^T(t+1)F(t)]/[1 + X^T(t+1)F(t)X(t+1)] +$$
$$+ Q(t) \tag{46}$$

The matrix Q(t) may be interpreted as the covariance of these parameter variations. It is used in the Kalman filter which will be presented in the next section.

In the following we give the code for a Fortran subroutine which performs a Least Squares Method with two Forgetting Factors for identification of system parameters.

```
    SUBROUTINE LSMFF(Y,N,PAR,F,X,E,F0,C,XTF,FX,FXXT,FXXTF)
c
c    LEAST SQUARES METHOD WITH TWO FORGETTING
c    FACTORS METHOD
c    THE SUBROUTINE MROEM PERFORMS THE
c    MODIFICATION OF THE PARAMETER ESTIMATES  PAR
c    FOR ONE SAMPLING TIME. A NEW CALL TO LSMFF MUST
c    BE MADE FOR EVERY NEW SAMPLING TIME
c
c    PAR   PARAMETER VECTOR
c    N     NUMBER OF PARAMETERS
c    Y     THE LAST OUTPUT VALUE
c    X     OBSERVATION VECTOR
c    E     PREDICTION ERROR
```

```
c    F      GAIN MATRIX WHICH MUST BE INITIALIZED

c    LA1    PARAMETER λ₁(.)

c    TF     THE TRACE OF THE MATRIX F

c    F0     THE DESIRED GAIN MATRIX TRACE

c    C  THE RATIO λ₁(.)/λ₂(.) WHICH IS KEPT CONSTANT

c    EQUAL TO C

c

DIMENSION PAR(N),X(N),XTF(N),FX(N),FXXT(N,N),FXXTF(N,N),
F(N,N)
      REAL LA1

c

c           compute prediction error

c

      E=Y

      DO 10 i=1,N

   10 E=E-PAR(i)*X(i)

c

      DO 11 i=1,N

      XTF(I)=0.0

      DO 11 j=1,N

   11 XTF(i)=XTF(i)+X(j)*F(j,i)

      XTFX=0.0

      DO  12 i=1,N

   12 XTFX=XTFX+XTF(i)*X(i)

      DO  13 i=1,N
```

```
      FX(i)=0.0
      DO  13 j=1,N
   13 FX(i)=FX(i)+F(i,j)*X(j)
      DO  14 i=1,N
      DO  14 j=1,N
   14 FXXT(i,j)=FX(i)*X(j)
      DO  15 i=1,N
      DO  15 i=1,N
      FXXTF(i,j)=0.0
      D0  15 k=1,N
   15 FXXTF(i,j)=FXXTF(i,j)+FXXT(i,k)*F(k,j)
      TF=0.0
      TFXXTF=0.0
      DO  16 i=1,N
      TF=TF+F(i,i)
   16 TFXXTF=TFXXF+FXXTF(i,i)
C
C              compute new parameter $\lambda_1(.)$
C
      LA1=(TF-TFXXTF/(C+XTFX))/TF
      DO 17 i=1,N
      DO 17 i=1,N
   17 F(i,j)=(F(i,j)-FXXTF(i,j)/(C+XTFX))/LA1
C
C              compute new parameter estimates
```

```
c

      DO 18 i=1,N
      DO 18 j=1,N
   18 PAR(i)=PAR(i)+F(i,j)*X(j)*E
      RETURN
      END
```

9. THE KALMAN FILTER ESTIMATOR

The Kalman filter was initially developed for optimal state reconstruction in the sense that the variance of the reconstruction error (the difference between the true parameters and their estimates) should be minimal (Anderson and Moore, 1979).

Let us consider the following state space representation:

$$x(t+1) = Ax(t) + Bu(t) + e(t)$$
$$y(t) = C^T x(t) + v(t) \tag{47}$$

where $x(t)$, $u(t)$ and $y(t)$ denote the state vector, the input and the output respectively.

The disturbances effects are assumed to be as follows:
$e(t)$ and $v(t)$ are sequences of zero mean, independent, equally distributed, normal random variables with covariances Q and R, respectively. It is also assumed that the sequences $e(t)$ and $v(t)$ are independent.

The Kalman filter associated with the above state representation (47) is given by:

$$x^*(t+1) = Ax^*(t)+Bu(t) + K(t)[y(t) - C^T x^*(t)]$$

$$K(t) = AF(t)C[C^T F(t)C + R]^{-1} \tag{48}$$

$$F(t+1) = A\{F(t) - [F(t)CC^T F(t)]/[R + C^T F(t)C]\}A^T + Q$$

$x^*(t+1)$ represents the state estimation given data up to and including time t.

It is worth mentioning that the standard least squares algorithm corresponds to the following design parameters (see equation 32):

$$x = \theta; \quad B = 0; \quad C = X(.); \quad R = 1 \text{ and } Q = 0$$

The innovations sequence $(y(t) - C^T x^*(t))$ represents the estimation error. This leads to

$$\theta^*_{t+1} = \theta^*_t + K(t)e_p(t)$$

$$K(t) = F(t)X(t+1)/[R + X^T(t+1)F(t)X(t+1)] \tag{49}$$

$$F(t+1) = F(t) + Q - F(t)X(t+1)X^T(t+1)F(t)/[R + X^T(t+1)F(t)X(t+1)]$$

The matrix $K(.)$ is called the Kalman gain. It weights the prediction error $e_p(.)$. Note that the random variables $\{e(t) \text{ and } v(t)\}$ have opposite effects on the Kalman estimator: a large measurement noise (R) decreases the covariance matrix $F(.)$, thereby decreasing the Kalman gain, whereas a large parameter disturbance (Q) increases the matrix $F(.)$ and thereby gain $K(.)$. The recursion defined by equations (49) minimizes the variance of the prediction error $e_p(.)$.

Comparing the gain matrix in (32) and (49) it can be seen that the recursive least squares algorithm is obtained if $R = 1$ and

$Q = 0$.

To improve the numerical stability of this estimation procedure, factorization techniques which will be also presented in this chapter can be used for the Kalman estimator implementation (Ljung and Söderström, (1983)).

The algorithm describing the updating of the vector parameter ø is realized by the Fortran subroutine KFIA.

```
SUBROUTINE KFIA(YS,N,PAR,E,GK,F,Q,R,FX,FXXT,FXXTF,XTF)
c
c        KALMAN FILTER IDENTIFICATION ALGORITHM
c
c     YS              THE LAST OUTPUT
c     PAR             THE VECTOR PARAMETERS
c     N               THE NUMBER OF PARAMETERS
c     E               THE PREDICTION ERROR
c     GK              THE KALMAN GAIN
c     Q,R             COVARIANCE MATRIX
c
DIMENSION PAR(N),GK(N),F(N,N),FX(N),FXXT(N,N),FXXTF(N,N),
XTF(N)
c
c                compute the prediction error
c
     E=YS
     DO 1 i=1,N
```

```
      E = E - X(i)*PAR(i)
      DO  2 i=1,N
      FX(i)=0.0
    2 FX(i)=FX(i)+F(i,j)*X(j)
      DO  3 i=1,N
      DO  3 j=1,N
    3 FXXT(i,j)=FX(i)*X(j)
      DO  4 i=1,N
      DO  4 j=1,N
      FXXTF(i,j)=0.0
      DO  4 k=1,N
    4 FXXTF(i,j)=FXXTF(i,j)+FXXT(i,k)*F(k,j)
      DO  5 i=1,N
      XTF(i)=0.0
      DO  5 j=1,N
    5 XTF(i)=XTF(i)+X(j)*F(j,i)
      XTFX=0.0
      DO  6 i=1,N
    6 XTFX=XTFX+XTF(i)*X(i)
c
c            compute the Kalman gain
c
      DO  7 i=1,N
    7 GK(i)=FX(i)/(R+XTFX)
      DO  8 i=1,N
c
```

```
c               covariance matrix adaptation
c
      DO  8 j=1,N
   8  F(i,j)=F(i,j)+Q(i,j)-FXXTF(i,j)/(R+XTFX)
c
c               compute new parameter estimates
c
      DO  9 i=1,N
   9  PAR(i)=PAR(i)+GK(i)*E
c
      RETURN
      END
```

Another choice of design parameters which has shown to be of practical importance is presented in the next section.

10. THE LEAKAGE EFFECT

The identification algorithms presented above require that the estimation error be zero mean and the available data be sufficiently rich (Johnstone and Anderson, 1982; Johnstone et al. 1982). Such a problem is due to the integral nature of the parameter estimator used.

When these assumptions are not fulfilled, parameter estimates may drift, leading to unstable behaviour of the underlying control or identification systems.

Several algorithm "fixes" have been proposed to deal with this problem. These include the dead zone algorithms (Egart, 1979; Samson, 1983), the parameter estimation contraction referred to as leakage (Ioannou and Kokotovic, 1982), the parameter estimation projection (Egart, 1979). Among all these solutions, the leakage-based algorithms have shown to be of practical significance. A suitable algorithm has been proposed by Irving (1985). It consists of choosing the design parameters in the Kalman filter as follows:

$$A = (1 - \beta)I \text{ with } 0 \le \beta < 1; \quad B = \beta; \quad u(t) = \theta^p$$

This leads to:

$$\theta^*_{t+1} = (1-\beta)\theta^*_t + \beta\theta^p + P(t)X(t+1)e(t) \tag{50}$$

$$F(t+1) = (1 - \beta I)F(t) + Q(t+1) - \frac{F(t)X(t+1)X^T(t+1)F(t)}{1 + X^T(t+1)F(t)X(t+1)} \tag{51}$$

The extra term $Q(.)$ prevents $F(.)$ from tending to zero. The design parameter β makes it possible to remove the integral nature of the standard parameter estimators. This provides a stable first order parameter estimator which corresponds to the following plant dynamics parametrization:

$$\theta^*_{t+1} = (1 - \beta)\theta^*_t + \beta\theta^p + e(t) \tag{52}$$

$$y(t) = X^T(t)\theta^*_t + v(t) \tag{53}$$

Equation (52) can be written more succinctly as:

$$\theta^{*}_{t+1} = \beta\theta^{p} / [1 - (1 - \beta)q^{-1}] \tag{54}$$

Equation (54) represents a stable process driven by θ^{p} and having θ^{*}_{t} as output. This process constant time is equal to: Tc = -T/Ln(Δ) (T is the sampling period). In the same way, it can be shown that the covariance matrix of the error prediction converges to:

$$\lim_{N \to \infty} P(t+1) = [(1 - \Delta)^{2}I]^{-1}Q(t+1) \tag{55}$$

The matrix Q(.) is generally as a diagonal matrix

$$Q(t+1) = [(1 - \Delta)2I]\Omega \tag{56}$$

where Ω is constant positive value.

Also from (44) we have

$$\lim_{N \to \infty} P(t+1) = \Omega.I \tag{57}$$

One important advantage of the leakage approach compared with other identification techniques is that a priori knowledge of the process parameters can easily be incorporated.

The Fortran code that performs the leakage effect method is given below.

```
SUBROUTINE LEAK(PAR,PARP,E,Y,N,X,XFF,FX,F,Q,FXXT,
                FXXTF,BETA)
c    LEAKAGE EFFECT IDENTIFICATION ALGORITHM
```

```fortran
c     DEVELOPPED BY E. IRVING
c
c     PAR    PARAMETER VECTOR
c     N         NUMBER  OF PARAMETERS
c     PARP   A PRIORI PARAMETERS (θᵖ)
c     X         OBSERVATION VECTOR
c     E         PREDICTION ERROR
c     F         GAIN MATRIX WHICH MUST BE INITIALIZED
c     Q         MATRIX Q  N.N
c     BETA   PARAMETER ß
DIMENSION PAR(N),PARP(N),X(N),XTF(N),FX(N),F(N,N),
DIMENSION Q(N,N),FXXT(N,N),FXXTF(N,N)
c
c            compute prediction error
c
      E=Y
      DO  1 i=1,N
    1   E=E-PAR(i)*X(i)
c
c            update the covariance matrix
c
      DO  2 i=1,N
      XTF(I)=0.0
      DO 11 j=1,N
    2   XTF(i)=XTF(i)+X(j)*F(j,i)
```

```
        XTFX=0.0
        DO  3 i=1,N
   3    XTFX=XTFX+XTF(i)*X(i)
        DO  4 i=1,N
        FX(i)=0.0
        DO  4 j=1,N
   4    FX(i)=FX(i)+F(i,j)*X(j)
        DO  5 i=1,N
        DO  5 j=1,N
   5    FXXT(i,j)=FX(i)*X(j)
        DO  6 i=1,N
        DO  6 i=1,N
        FXXTF(i,j)=0.0
        D0  6 k=1,N
   6    FXXTF(i,j)=FXXTF(i,j)+FXXT(i,k)*F(k,j)
        DO 7 i=1,N
        DO 7 j=1,N
   7    F(i,j)=F(i,j)+Q(i,j)-FXXTF(i,j)/(1.0+XTFX)
c
c           update new parameter estimates
c
        DO  8  i=1,N
        DO  8  j=1,N
   8    PAR(i)=(1.0-BETA)*PAR(i)+BETA*PARP(i)+F(i,j)*X(j)*E
        RETURN
        END
```

11. THE NORMALIZATION AND PROJECTION IDENTIFICATION ALGORITHM

The algorithm "fixes" which avoid parameter drift are based on the bounded disturbance assumption. Such an assumption is not generally satisfied in the presence of unmodelled dynamics. Indeed the disturbance effects depend on the input and output sequences in this case. To overcome this problem, a judicious solution has been proposed by Praly (1984). It consists of normalizing the data before entering the parameter adaptation algorithm.

Let us introduce the normalization factor $r(t)$ which will be used to normalize the observation vector $X(t)$ and the prediction error $e(t)$. The value of $r(.)$ can be updated, for instance by using

$$r(t+1) = \mu.r(t) + \max[X^T(t)X(t), r_0] \tag{58}$$

Equation (49) is a first order system with the pole μ. The parameter μ determines how fast the factor r will approach $\max[X^T(.)X(.), r_0]$. More precisely, the unmodelled dynamics should be outside of the normalization filter bandwidth. The parameter r_0 prevents $r(.)$ from tending to zero and may be viewed as a bound on the external disturbances.

The normalization features can be used with any algorithm "fixes". In Praly (1983), this concept has been used together with the parameters projection idea to ensure the global stability of pole placement adaptive controller. The involved parameter adaptation algorithm is given by:

$$\theta'_{t+1} = \theta^*_t + g(t)F(t)X_n(t+1)e_n(t) \tag{59}$$

$$\theta^*_{t+1} = \begin{cases} \theta'_{t+1} & \text{if } \| \theta'_{t+1} - \theta_c \| \leq R \\ \\ \theta_c + R\{\theta'_{t+1} - \theta_c\}/\| \theta'_{t+1} - \theta_c \| \end{cases} \tag{60}$$

$$g(t+1) = 1/[\beta(t+1) + X_n^T(t)F(t)X_n(t+1)] \tag{61}$$

$$F'(t+1) = F(t) - g(t+1)F(t)X_n(t+1)X_n^T(t+1)F(t) \tag{62}$$

$$F(t+1) = [1 - \beta_4/\beta_3]F'(t+1) + \beta_4.I \tag{63}$$

with

$$X_n(t) = X(t)/\sqrt{r(t)} \quad , \quad e_n(t) = e(t)/\sqrt{r(t)} \tag{64}$$

θ_c and R denote the center and the radius of the projection sphere $S(\theta_c, R)$.

The constants β_3 and β_4 prevent the adaptation matrix gain from going to zero; i.e.,

$$\beta_4.I \leq F(.) \leq \beta_3.I \tag{65}$$

Notice that this algorithm needs a priori knowledge about the plant dynamics in terms of the projection sphere $S(\theta_c, R)$. This knowledge is none the less very easy to get.

One difficulty in using this algorithm in practical situations is its slow rate of convergence. The normalization and projection identification algorithm can converge many times faster if equation (49), which is related to the normalization factor, is modified as follows:

$$r(t+1) = \mu r(t) + (1 - \mu)\max[X^T(t+1)X(t+1), r_0] \tag{66}$$

A Fortran mechanization of this algorithm is presented here.

```
c    NORMALZATION AND PROJECTION IDENTIFICATION
c    ALGORITHM DEVELOPPED BY L. PRALY
c
SUBROUTINE NPIA(Y,N,PAR,F,X,E,ROØ,DD,MU,BETA,BETA3,
 ,BETA4,XN,,XNTF,FXN,FXNXNT,FXNXNTF)
c
c    PAR   PARAMETER VECTOR
c    N       NUMBER OF PARAMETERS
c    Y       THE LAST OUTPUT VALUE
c    X       OBSERVATION VECTOR
c    E       PREDICTION ERROR
c    F        GAIN MATRIX WHICH MUST BE INITIALIZED
c    DD     RADIUS SPHERE
c    XN    THE NORMALIZED VECTOR X
c    ROØ   THE PARAMETER r0
c
DIMENSION PAR(N),X(N),XN(N),XNTF(N),FXN(N)
DIMENSION F(N,N), FXNXNT(N,N),FXNXNTF(N,N)
```

```fortran
      REAL MU
c
      XNN=0.0
      DO  1 i=1,N
    1  XNN=XNN+X(i)*X(i)
      XNN=SQRT(XNN)
      RXX=ROØ
      IF (XNN.GT.ROØ)  RXX=XNN
      RO=MU*RO+RXX
c        compute the prediction error
      E=Y
      DO  2 i=1,N
    2 E=E-X(i)*PAR(i)
c        compute the normalized values of the prediction error
c              and the observation vector
      DO  3 i=1,N
      EN=E/SQRT(RO)
    3 XN(i)=X(i)/SQRT(RO)
      DO  4 i=1,N
      XNTF(i)=0.0
      DO  4 j=1,N
    4 XNTF(i)=XNTF(i)+XN(j)*F(j,i)
      XNTFXN=0.0
      DO  5 i=1,N
    5 XNTFXN=XNTFXN+XNTF(i)*XN(i)
      G=1.0/(BETA+XNTFXN)
```

```
c         compute new parameter estimates
      DO  6 i=1,N
      DO  6 j=1,N
    6 PAR(i)=PAR(i)+G*F(i,j)*XN(j)*EN
      PARN=0.0
      DO  7 i=1,N
    7 PARN=PARN+PAR(i)*PAR(i)
      PARN=SQRT(PARN)
      IF(PARN.GT.DD)    GOTO 8
      DO  9 i=1,N
    9 PAR(i)=PAR(i)*DD/PARN
    8 CONTINUE
c         compute new gain matrix
      DO 10 i=1,N
      FXN(i)=0.0
      DO 10 j=1,N
   10 FXN(i)=FXN(i)+F(i,j)*XN(j)
      DO 11 i=1,N
      DO 11 j=1,N
   11 FXNXNT(i;j)=FXN(i)*XN(j)
      DO 12 i=1,N
      DO 12 j=1,N
      FXNXNTF(i,j)=0.0
      DO 12 k=1,N
   12 FXNXNTF(i,j)=FXNXNTF(i,j)+FXNXNT(i,k)*F(k,j)
      DO 13 i=1,N
```

```
        DO  13 j=1,N
        F(i,j)=F(i,j)+(1.0-BETA4/BETA3)(F(i,j)-G*FXNXNTF(i,j))
        IF (i.EQ.j)     F(i,j)=F(i,j)+BETA4
  13  CONTINUE
        RETURN
        END
```

Several methods can be used to perform the recursions appearing in on-line parameter estimation procedures. The most crucial problem to be taken into consideration is the propagation of the round-off errors that will deteriorate the estimation and consequently the performance of the associated control algorithm in adaptive contex.

Two factorization methods commonly used in practice will be presented in the next section of this chapter.

12. FACTORIZATION METHODS

In all identification schemes, it seems easy, at first sight, to implement the recursion associated to the covariance matrix $P(t)$ which is often called the Riccati equation (see the section concerned the L.Q.G. controller design in the following chapter) or the Kalman equation (which is similar to that related to the solution of filtering and state estimation problems provided by the Kalman filter). Unfortunately, the Ricatti equation has unsatisfactory numerical performances. The main reason is that $F(t)$ may be essentially computed by successive substractions $(F(t + 1) = F(t) - $ [something]). This matrix corresponds to the second condition of optimality (in the mean squares sense). It must be non-negative definite. Owing the round-off errors, which are the consequence of using computer systems, the covariance matrix may lose this property (the matrix $F(t)$ becomes ill-conditioned) and so a problem of numerical instability can occur. To avoid this, factorization methods can be implemented to update, not directly the covariance matrix $F(t)$, but factors appearing in a certain decomposition (factorization) of this matrix. In this section we shall be concerned with the popular factorization methods: square root (called also Cholesky decomposition) and U/D factorization methods (Bierman, 1977; Favier, 1987).

The U/D factorization method consists of decomposing the covariance matrix as follows:

$$F(t) = U(t)D(t)U^T(t) \tag{67}$$

where $U(t)$ is an upper triangular matrix with all diagonal elements are equal to unity, and $D(t)$ is a diagonal matrix, while the square-root factorization leds to the following decomposition:

$$F(t) = [F(t)]^{1/2} [F^T(t)]^{1/2} \tag{68}$$

where the nonsingular matrix $[F(t)]^{1/2}$ is the square root of $F(t)$

To make the presentation clear, only the sqare-root factorization method will detailed in the following.

As in Favier (1987), let us introduce the following notation to describe the weighted least squares estimation algorithm.

$$\Sigma(t) = \lambda^{-1}(t) + X^T(t)F(t-1)X(t) \tag{69}$$

$$K(t) = F(t-1)X(t)\Sigma^{-1}(t) \tag{70}$$

$$F(t) = [F(t-1) - K(t)\,\Sigma(t)K^T(t)]/\lambda(t) \tag{71}$$

$$\theta(t) = \theta(t-1) + K(t)\,[y(t) - X^T(t)\theta(t-1)] \tag{72}$$

A suitable alternative to carrying out a factorization of the matrix F(t) consists of writing equations (69 - 71) in a compact matrix form:

$$M(t) = N(t) \tag{73}$$

with

$$M(t) = \begin{bmatrix} \Sigma(t) & \Sigma(t)K(t)^T \\ K(t)\Sigma(t) & \lambda(t)\,F(t) + K(t)\Sigma(t)K(t)^T \end{bmatrix}$$

$$N(t) = \begin{bmatrix} \lambda(t)^{-1} + X(t)^T F(t-1)X(t) & X(t)^T F(t-1) \\ F(t-1)X(t) & F(t-1) \end{bmatrix}$$

These matrices can be factorized as follows:

$$M(t) = [M_S(t)]^{1/2}[M_S(t)^T]^{1/2} \tag{74}$$

$$N(t) = [N_S(t)]^{1/2}[N_S(t)^T]^{1/2} \tag{75}$$

The subscript s denotes the square-root. The matrix $M_S(t)$ and $N_S(t)$ given by:

$$M_S(t)^{1/2} = \begin{bmatrix} \Sigma(t)^{1/2} & 0 \\ K(t)\Sigma(t)^{1/2} & \lambda(t)^{1/2}\,F(t)^{1/2} \end{bmatrix}$$

$$N_S(t)^{1/2} = \begin{bmatrix} \lambda(t)^{-1/2} & X(t)^T F(t\text{-}1)^{1/2} \\ 0 & F(t\text{-}1)^{1/2} \end{bmatrix}$$

The matrices M(t) and N(t) are equal, as is the case with the matrices $M_S(t)$ and $N_S(t)$. Using any orthogonalization method such as those of Householder , Gram-Schmidt, etc. (Golub and Von Loan, 1983; Franklin, 1968), the factorization problem can be solved as follows:

$$N_S(t) = Q^* T \tag{76}$$

where the matrix Q^* is defined as follows:

$$N_S(t)^{1/2} = \begin{bmatrix} \lambda(t)^{-1/2} & X(t)^T F(t\text{-}1)^{1/2} \\ 0 & F(t\text{-}1)^{1/2} \end{bmatrix} = \begin{bmatrix} a & 0 \\ b & C \end{bmatrix} T$$

The matrix T is orthogonal (a square matrix $T \epsilon R^{n+1.n+1}$ is said to be orthogonal if $T^T T = I$, where $R^{n+1.n+1}$ is Euclidean (n+1).(n+1) dimensional space and I is the unit matrix).

Comparison of equations (75 - 76) provides directly the square-root of the covariance matrix F(t)

$$[F(t)]^{1/2} = [\lambda(t)]^{-1/2} C , \quad K = b/a \tag{77}$$

The gain K is usuallay called the Kalman gain vector.

Finally, to give an overview of the two appropriate factorization methods (the square-root and U/D method, which are also called Peterka's and Bierman and Thorton methods respectively), we give below a listing of the U/D method. The performances of these two methods are comparable. The U/D method requires no root extraction and explicitly provides parameter variance which can be used in an estimation supervizor procedure for diagnostic purposes. In the following subroutine we shall also include an adaptation of the forgetting factor to maintain constant the trace of the covariance matrix F(t), and consider the general case when the covariance matrix Q of the random vector associated to the parameters is taken into consideration as in the Kalman filter estimator. A procedure for monitoring the elements of the matrix D to ensure a lower bound on the estimator gain matrix is also introduced. The upper bound is ensured by the constant trace procedure.

```
      SUBROUTINE UDU(NP,U,D,X,IN,AK,C,TR,TD,PAR,ITRA,AL1,
     ,AL2,FORGET,W,V,DP,IN,VD,YS)
c     PAR          THE VECTOR PARAMETERS
c     NP           THE NUMBER OF PARAMETERS
c     AK           THE KALMAN GAIN
c     U            AN UPPER TRIANGULAR MATRIX
c     D            A DIAGONAL MATRIX, D = αI, α > 0
c     YS           THE LAST OUTPUT
c     X            THE OBSERVATION VECTOR
c     TR           THE TRACE OF THE COVARIANCE MATRIX
c     TD           THE DESIRED TRACE
c     IN           THE DIMENSION OF THE COVARIANCE
```

```
c                     MATRIX Q IS NP.IN
c     VD              THE DESIRED VALUE OF THE ELEMENTS
c                     OF THE MATRIX D
c     ITRA            ITRA = 1 MAINTAIN CONSTANT THE TRACE
c                     OF THE COVARIANCE MATRIX
c     AL1             THE FACTOR λ₁(t)
c     AL2             THE FACTOR λ₂(t)
c     FORGET          FORGETTING FACTOR
c
      DIMENSION PAR(1), D(1), U(1,1), AK(1), X(1), W(1,1),  V(1),
     ,DP(1), F(1)
c
      ER=YS
      DO 10 i=1,NP
 10   ER=ER-PAR(i)*X(i)
      DO 11 i=1,NP
      F(i)=0.
      DO 11 j=1,i
 11   F(i)=F(i)+U(i,j)*X(j)
      Do 12 i=1,NP
 12   G(i)=D(i)*F(i)
      IF(ITRA.EQ.1) THEN
      BETA0=FORGET
      ELSE
      BETA0=AL1/AL2
```

```
      ENDIF
      DO 13 j=1,NP
      BETA1=BETA0+F(j)*G(j)
      D(j)=BETA0*D(j)/BETA1
      V(j)=G(j)
      AMU=-F(j)/BETA0
      IF(j.NE.1) THEN
      DO 14 i=1,j-1
      PAU=U(i,j)
      U(i,j)=U(i,j)+AMU*V(i)
      V(i)=V(i)+PAU*V(j)
14    CONTINUE
      ENDIF
      BETA0=BETA1
13    CONTINUE
      IF(ITRA.EQ.1) THEN
      BETA1+BETA1-FORGET+1
      ELSE
      BETA1+BETA1-AL1/AL2+1
      ENDIF
      DO 15 i=1,NP
15    AK(i)=V(i)/BETA1
      DEN=BETA1
      IF(ITRA.EQ.1) THEN
      TR=0.
      DO 16 i=1,NP
```

```
      DO 16 k=1,NP
      TR=TR+U(i,k)*D(k)*U(i,k)
16    CONTINUE
      AL1+TR/TD
      AL2+ALI/FORGET
      DO 17 i=1,NP
      D(i)=D(i)/AL1
      IF(D(i).LT.VD) THEN
      D(i)=VD
      ENDIF
17    CONTINUE
      ELSE
      DO 18 i=1,NP
      D(i)=D(i)/AL1
      IF(D(i).LT.VD) THEN
      D(i)=VD
      ENDIF
18    CONTINUE
      IF(IN.EQ.0) THEN
      ELSE
      DO 19 i=1,NP+IN
      DO 19 j=1,NP
      IF(i.LE.NP) THEN
      W(i,j)=U(j,i)
      ELSE
      W(i,j)=C(j,i-NP)
```

```fortran
      ENDIF
19    CONTINUE
      DO 20 i=1,NP
20    DP(i)=D(i)
      DO 21 i=NP+1,NP+IN
21    DP(i)=1.0
      I=NP
17    IF(I.LT.2) GO TO 22
      D(I)=0.
      DO 23 k=1,NP+IN
23    D(I)=D(I)+W(k,I)*W(k,I)*DP(k)
      DO 24 i=1,I-1
      U(i,I)=0.
      DO 25 k=1,NP+IN
25    U(i,I)=U(i,I)+W(k,i)*W(k,I)*DP(k)/D(I)
      DO 24 k=1,NP+IN
      W(k,i)=W(k,i)-U(i,I)*W(k,I)
24    CONTINUE
      I=I-1
      Go TO 17
22    CONTINUE
      D(1)=0.
      DO 28 k=1,NP+IN
28    D(1)=D(1)+W(k,1)*DP(k)*W(k,1)
      ENDIF
      ENDIF
```

```
      TR=0.
      DO 29 i=1,NP
      DO 29 k=1,NP
      TR=TR+U(i,k)*D(k)*U(i,k)
29    CONTINUE
      DO 30 i=1,NP
      PAR(i)=PAR(i)+AK(i).ER
30    CONTINUE
      RETURN
      END
```

<u>Remark:</u>

These factorization techniques can also be used to solve the Riccati equation involved in the Linear Quadratic Gaussian controller design which shall be developed in the next chapter.

13. CHOICE OF IDENTIFICATION PARAMETERS

The application of identification techniques involves the choice of a number of parameters such as the model order, the sampling time, the delay,etc. When appropriate prior knowledge about the process dynamics is available, it must be used for identification purposes.

A lot of papers are devoted to this problem. There exists no general method and most of the assumptions made to derive these methods are not coherent with real-life situations. From our viewpoint, we think a lot of information can be obtained about the process dynamics from its transient response (step or impulse response). First and second order linear systems with time delay are the most commonly used approximate models, because on the one hand they are relatively simple and have been extensively studied, and on the other hand they provide a good approximation of the nonlinear behaviour of real processes to be controlled. A lot of processes can be described by a second order model: furnace (Najim, 1982; Najim and Muratet, 1983; Najim et al., 1985), reactor (Lee and Won-Kyoo, 1985; Koutchoukali et al., 1986), absorber-desorbers (Kershenbaum and Fortescue, 1981), crystallization plants (Kershenbaum and Fortescue, 1981), digesters (Belanger et al. 1986), distillation columns (Dahlqvist, 1981), liquid-liquid extraction columns (Najim and Muratet, 1987; Najim et al., 1986b).

14. PRACTICAL ASPECTS OF ESTIMATION ALGORITHMS

In this section we shall be concerned with practical implementation of the estimation algorithm and with the measures which enhance robustness and numerical stability. It is worth noticing that all or part of the modifications of the basic recursive least squares algorithm introduced herein will be useful in order to make the estimation procedure capable of dealing with a wide range of practical situations. This is the case for almost all the chemical plants considered at the end of this chapter.

14.1. FACTORIZATION

When computations are performed on a digital computer, each arithmetic operation is generally affected by roundoff error. This error arises because the machine hardware can only represent a subset of the real numbers (example: the basic M6800 EXORciser has a characteristics Word size of 8 bits). In multiplication, a double-length result is produced and the machine must reduce the number of bits in the product to fit it into the standard word size. This is often done by ignoring the least significant half of the product. The some problem occurs with division.

It should be noticed, firstly, that the covariance matrix $P(t)$ must remain positive definite. However, even if the initial matrix $P(0)$ satisfies the second order condition of optimality (least squares minimization), the positive definiteness of $P(t)$ can be lost (owing to numerical round-off errors). Therefore it is more advisable to update the estimator in a factorized form which guarantees that $P(t)$ remains positive definite and that the round-off errors, which are inherent to the use of limited word length in computation, do not affect the solution significantly. The most popular method is the UD factorization which is based on the decomposition of $P(t)$ (as in

Bierman, 1977):

$$P(t) = U(t)D(t)U^T(t) \tag{78}$$

where the factors U and D are, respectively, an unitary upper triangular matrix and a diagonal matrix. To ensure a lower boundary on the estimator gain it is sufficient in this case to monitor the elements d_i of the diagonal matrix D(t). This is referred to as the regularization of the algorithm (Ljung and Söderström, 1983), i.e.,

$$d_i(t) = \max\{d_0, d_i(t)\} \tag{79}$$

where d_0 is a regularizing constant.

14.2. SIGNAL CONDITIONING

From a practical engineering point of view, we should point out that signal conditioning is important to obtain good data for either identification or control purposes. There will be no harm in performing signal filtering in order to obtain smooth measurements . Practically almost all industrial sensors have some kind of filter.

All signals are presumed to be disturbed. It is not desirable to accept the measured value y(t) as the true value of the output or measured perturbation signals. It is advisable to smooth the current values with past conditioned values in order to eliminate or to reduce the effects of noise.

An analog or digital filter may be used. Generally an analog filter are composed of a resistance and a condenser, as shown in figure 9. The transfer function of the analog filter shown in figure 9 has a transfer function of [1/(1 + RC s)], where s is the Laplace operator.

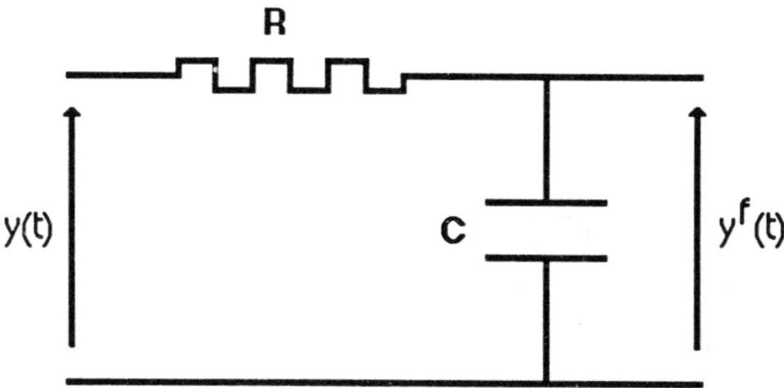

Figure 9. First order analog filter

A Fortran program which performs digital filtering is given below.

```
      SUBROUTINE FILTER(F,NF,G,NG,Y,YF,N)
c     DATA FILTERING
c            yf(t) = [F(q-1)/G(q-1)]y(t)
c     Y      VECTOR OF DATA TO BE FILTERED [y(t), y(t-1), ...]
c     YF     VECTOR OF FILTERED DATA [yf(t), yf(t-1), ...]
c     NY     DIMENSION OF VECTOR Y AND YF
c     F      COEFFICIENTS OF THE POLYNOMIAL F(q-1)
c     G      COEFFICIENTS OF THE POLYNOMIAL G(q-1)
c     NF     DEGREE OF THE POLYNOMIAL F(q-1)
```

```
c    NG      DEGREE OF THE POLYNOMIAL G(q-1)
c
     DIMENSION F(20),G(20),Y(20), YF(20)
     FILT=0.
     DO 1 i=2,NF+1
1    FILT=FILT -Y(i)*F(i)
     DO 2 i=1,NG+1
2    FILLT= FILT+G(i)*YF(i)
     DO 3 i=NY,2,-1
3    YF(i)=YF(i-1)
     YF(1)=FILT
     RETURN
     END
```

Another simple approach consists of performing the so called single exponential smoothing described by:

$$y^f(t) = \lambda y(t) + \beta y^f(t) \tag{80}$$

For $\beta = 1 - \lambda$, the response of the system associated to equation (80) has an exponential profile. The smoothing factor lies in the range [0, 1]. The filtered output is calculated by adding a certain fraction of the measured output and some fraction of the previous filtered output. When the confidence affected to the measured output $y(t)$ is small, the factor λ may be chosen small. This is the case when the measured signal is highly disturbed (high noise level). When the measured signal is not corrupted by noise the factor λ may be chosen close to one.

Equation (80) has some connections with the stochastic approximation approach, also called the stochastic gradient method.

Let us consider a noisy measured variable $y(t)$. The problem to be solved can be stated as follows: Find the mean value value of the stochastic variable $y(t)$. The solution of this problem has been given by Robbins Monro (1951) as follows:

$$y_m(t) = y_m(t-1) + \gamma_n[y(t) - y_m(t-1)] \tag{81}$$

The sequence γ_n verifies the following conditions:

$$\Sigma \gamma_n = \infty, \; \Sigma \gamma_n^2 = \infty \; n = 1, \ldots, \infty \tag{82}$$

A common choice of γ_n is : $\gamma_n = 1/n$. With this choice, equation (81) leads to:

$$y_m(t) = [\sum_{i=1}^{n} y(i)]/n \tag{83}$$

which corresponds to the expression commonly used for estimating the mean value of a random variable.

Nottice that for $\lambda = 1$ and $\beta = \gamma_n$ we obtain the expression (80).

The unmodelled dynamics (neglected dynamics, for example modelling a distributed parameter system which has infinite states with a lumped system wihich has finite states) result from the use of an input-output linear model to represent a more complex though nonlinear actual process. This is the case in particular for most chemical processes. Therefore parts of the process dynamics are neglected and these introduce extra modelling errors which are not necessarily bounded. It is therefore advisable to perform normalization of the input-ouput data before they are processed by the algorithm. A suitable normalization operator for the regressor and the prediction errors in the estimation algorithm can be defined as follows:

$$x_n(t) = x(t) \, / \sqrt{s(t)} \tag{84}$$

where $s(t)$ is a normalizing function to be computed using the input-output measurements. The following filter has been extensively used in the literature (Praly, (1984)):

$$s(t) = \mu s(t-1) + (1-\mu) \max\{s_0, \Phi_n^T(t-1)\Phi_n(t-1)\} \tag{85}$$

The constant time of the filter must be chosen such that the non-considered time constant in the model (neglected dynamics) may be filtered. The normalization ensures an upper bound of the matrix $F(t)$.

14.3. INFORMATION MEASURE

Least squares type algorithms with exponential forgetting will be unsatisfactory if there are intervals of little or no plant excitation. It is therefore advisable to compute a measure of information at every sampling interval using the input-output data, and to implement a decision rule to take into consideration how the current information can improve the estimation process. The following function can be used to measure how much the incoming information differs from previous information (M'Saad et al. (1986)):

$$\beta(t) = \lambda(t)\Phi_n{}^T(t-1)F(t-1)\Phi_n(t-1) \tag{86}$$

If $\beta(t)$ is below a certain value, one can either freeze the parameters at their previous values, or use other identification techniques based either on the leakage effect or on the projection procedure, (Praly, (1984)). Another technique consists of introducing a persistent exciting auxiliary input signal, but this solution is not accepted by process engineers in most practical situations.

14.4. DEAD TIME ESTIMATION

Several techniques for estimating the dead time k are discussed in this section. It is worth remarking that many adaptive control algorithms are sensitive to the time delay parameter. Therefore on-line estimation of this parameter will be of considerable interest, particularly in the case of plants with varying time delay.

In deterministic cases, the most popular technique uses the step response to identify the time delay. It is based on the measurement of the interval separating t he time when a step is applied to the input and the time when the output varies from its previous steady state. It is quite difficult to carry out this procedure if the considered plant is disturbed   by high level noise. In the adaptive control context, one way to deal with this problem consists of extending the B polynomial by a number of extra parameters to incorporate the maximum expected time delay, and finding the first non-zero b_i parameters. The difficulty with this method is due to the fact that the estimated b_i- parameters will never be exactly zero, and the problem is being translated to determine the threshold for parameters that are practically zero. This method does not yield an exact estimation of the time delay, and may drastically increase the number of parameters to be estimated.

Another procedure for estimating the time delay uses the multimodel approach. In this case several models with different values for the time delay are simultaneously identified. A decision rule is incorporated to determine the best structure.

In De Keyser (1986), three candidate structures are used (figure 10). The first indicates that the time delay should be increased, and the third indicates that the time delay should be decreased.

These models are excited by a white noise signal and two error sequences are computed and their variances are compared to select the best candidate structure.

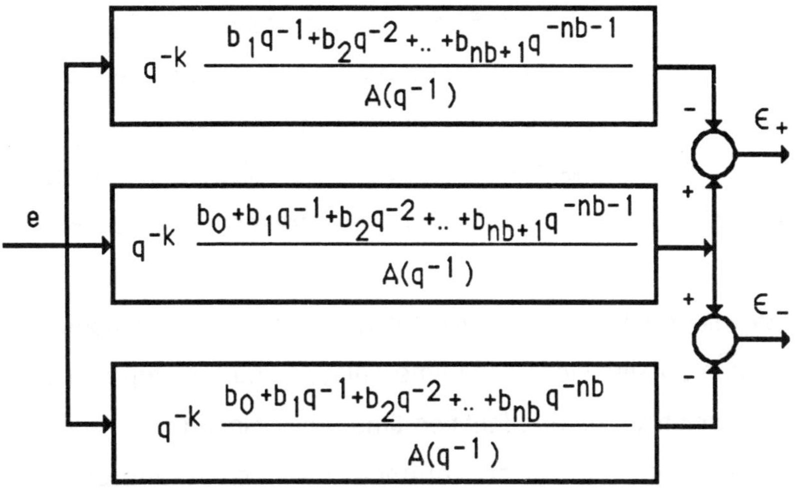

Figure 10. On-line estimation of the time delay using three models

The model structure which lead to the smallest values of the previous computed variances is selected. These variance calculations lead to:

$$E\{\epsilon_+\}/E\{\epsilon_-\} = b_0/[1 + (b_{nb} + 1)^2] \qquad (87)$$

For practical implementation, De Keyser (1986) suggest the following rules of thumb:

If $|b_0| > 5|b_{nb} + 1|$ then decrease the dead time

If $|b_{nb} + 1| > 5|b_0|$ then increase the dead time

Another approach is based on the comparison between the variances of the prediction errors of each structure.The time delay associated with the model which leads to the smallest prediction error variance is selected. Simulations of these techniques have demonstrated that these methods perform well when the considered time delay is close to the real one. Nevertheless, we have noticed that the method based on prediction error variance generally gives better results than that presented in De Keyser (1986).

We suggest a more advisable approach which combines an approximate estimation of the time delay, using the step response method which yields an identified time delay close to the real one, and the prediction error variance method to improve this estimation. This approach is quite useful in the case of systems with variable dead-time.

Notice that identification schemes are generally associated with control strategies. In this case, the computation time associated with identification and control can modify the dead time associated with the process under consideration.

In practice, two solutions may be adopted for on-line identification and control implementation using real-time computer systems.

At time t, the channel corresponding to the output $y(t)$ to be measured is given, and the computer program related to data aquisition runs. Then, after a certain time which will be noted t_a, the numerical value associated to the continuous variable $y(t)$ by the analog-digital converter is available at time $(t + t_a)$. At this point, the identification program can be started, followed by the control program.

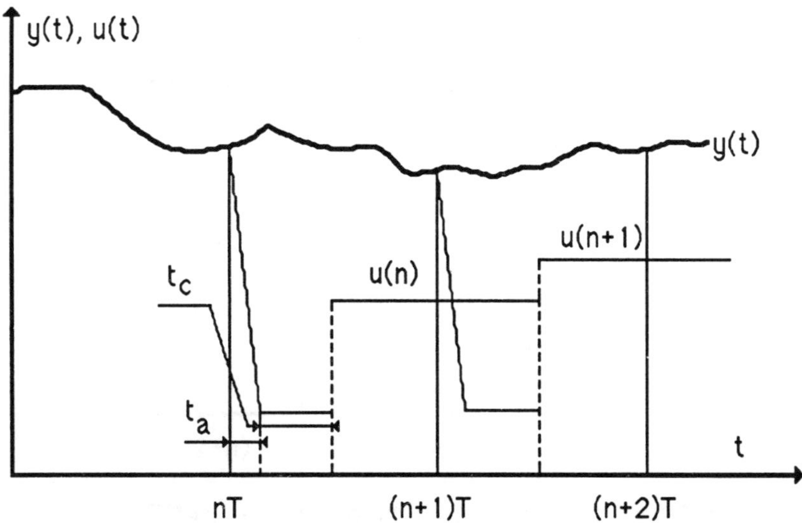

Figure 11. Input and output timing (delay $= kT + t_a + t_c$)

This computation takes a certain time which can vary from one sampling period to other. Let t_c be the time associated with identification and control programs.

At time $[t + t_a + t_c]$, two solutions can be considered:

-1. the new value $n(t + 1)$ of the control can be applied to the process at time

$[t + t_a + t_c]$ as shown in figure 11.

-2. The control $u(t + 1)$ may be applied at time $t + T$ as shown in figure 12.

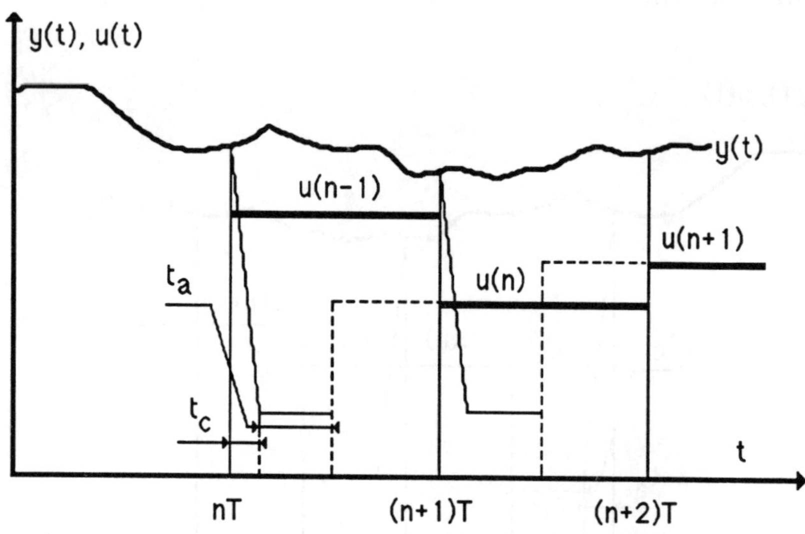

Figure 12. Input output timing (delay = (k+1)T)

The first solution introduces a supplementary time-varying dead time (generally small) equal to the global computation time of the control algorithm $(t_a + t_c)$.

The second solution leads to a constant extra time delay, equal to the sampling rate.

14.5. MODEL ORDER AND SAMPLING PERIOD SELECTION

The most important problem to be solved in real-time identification using recursive parameter estimation algorithms is the specification a priori of the order in the mathematical model. Many methods have been developed dealing with model validation and estimation of the structural parameters (Stoica et al. (1986)). These global methods are useful when off-line identification can be performed. As we are dealing with real-time identification, implementation of these methods is cumbersome both for computation time and for memory requirements. There are, however, other methods that may permit to proxy and choose an appropriate model order value, as for instance step response analysis. It should be noticed, however, that most practical applications of control theory generally use a second order input-output representation.

The first step in design of digital control system is the selection of an appropriate t sampling rate. It has to be chosen so that the continuous process response would be correctly recovered from the discrete data. This parameter is commonly specified using the following rule of thumb:

$$T_s/10 \leq T \leq T_s/4 \tag{88}$$

where T_s is the settling time of the plant under consideration.

Relation (88) means that the transient response chart of a given process can be recovered by a number of points between 4 and 10, which constitute a good representation of the continuous behaviour of the process under consideration.

Another way to select the sampling period may be stated as follows: the sampling rate must also be at least twice the desired closed-loop bandwidth, and possibly as much as 20 times the system closed-loop bandwidth, depending on the performance requirements of the system.

Another problem which must be taken into consideration is the time delay of the sensors. The sampling period must be very much less than the sensor dead time.

In the case of multivariables, it happens that certain parameters have faster time evolution than others. In this case, it is necessary to use multiple sampling periods (Najim and Al Khani, 1988).

14.6. ESTIMATION PROCEDURE INITIALIZATION

Good initialization of the identification algorithm is important to step down the transient phase, particularly when used in an adaptive context. It is advisable to take into consideration any available a priori information on the identified process. If no a priori information is available, an off-line identification may be carried out using a pseudo-random binary sequence (prbs) control signal, with the initial values of the parameters set to zero and the estimator gain set to

$F(0) = \alpha I$, where I is the identity matrix and $\alpha > 0$.

The value of α may be chosen high enough to allow fast adaptation of the parameters, while a small value of α leads to a slower adaptation.

15. CONCLUSION

In this chapter a series of different kinds of process models is given, and different approaches to the synthesis of robust real-time algorithms for the identification of dynamic systems are presented.

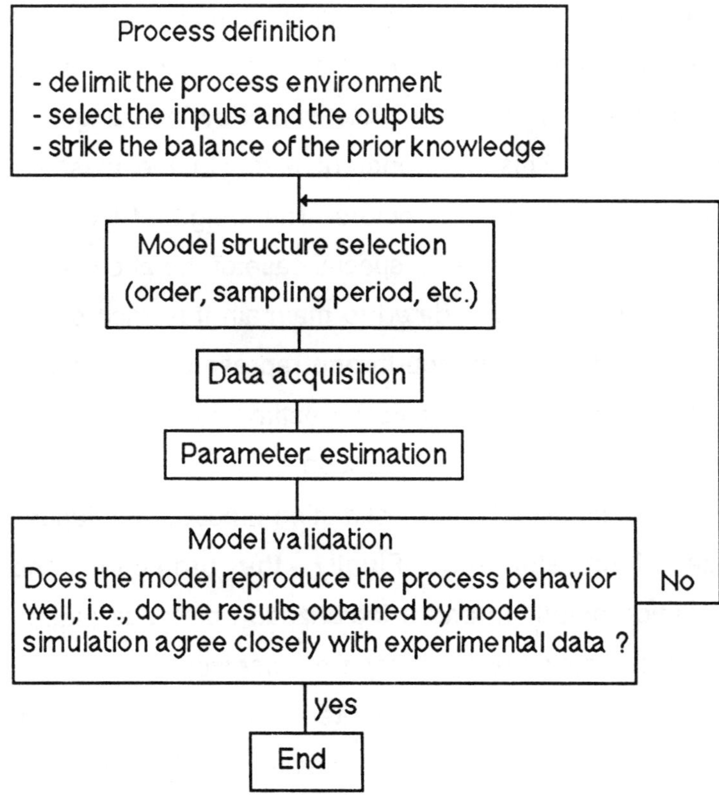

Figure 13. Flowsheet of model buiding

One approach is based on the derivation of recursive estimation procedures from the least squares criterion. To deal with biased estimates, the generalized least squares method is presented.

The procedure for building models using identification techniques can be summarized in the following flowsheet (figure 13):

To discount old data, a forgetting factor have been introduced. The second approach follows from the modification of the recursion related to the covariance matrix by including a second weighting factor. This development leads to the least squares method with two forgetting factors. The weighted least squares algorithm can be derived as a special case of the above algorithm. The forgetting factor is updated to maintain the trace of the gain matrix constant, while the fourth one represent the application of the well known Kalman filter to the estimation problem. The fifth method is related to the introduction of prior knowledge on the process to be identified in the estimation procedure. It leads to the leakage effect algorithm. Finally, the projection and the normalization algorithm and a factorization methods (square-root and U/D factorization techniques) are presented.

Parameter estimation using filtered data is more resistant to low frequency disturbances than estimation without filtering.

We have tried to insist on the algorithmical and practical aspects rather than on the mathematical concepts (a wide variety of methods : Liapunov functions, hyperstability, positive real transfer functions, ordinary differential equation, martingales are used for convergence and stability purposes). We have also given special consideration to the computer programs. For simplicity we have not presented a general program for identification purposes.

Arbitrary sets of model parameters and high values for the covariance matrix can always be used to start the recursive estimation scheme. Neither of these make any significant difference to the behaviour of the algorithm.

2

ADAPTIVE CONTROL

1. INTRODUCTION

In the process control industry, the analog P.I.D.-controller has been used for a long time to provide proportional, integral and derivative control action. The integral term is switched off for certain types of regulator loop to enhance the stability.

Many rules exist for adjusting continuous (Ziegler and Nichols, 1943; Naslin, 1963) and discrete P.I.D.-regulators (Wittenmark, 1979). Furthermore, frequent readjustment of the controller constants may be required if process operating conditions change significantly under set point variations or because of disturbances like variations of the properties of the feed, the catalyst, the extractor characteristics, etc. This in turn leads to manual control. It is desirable to obtain automatic tuning of P.I.D.-controllers in order to achieve an acceptable level of performance of the control system.

To implement such self-adjusting regulators, a digital computer must be used, and self-tuning PI.D.-controllers work well only for a reduced class of systems (free-time delay, etc.). Technological advances in large-scale integration are increasing the capabilities of hardware techniques. These advances offer the engineer the prospect of implementing adaptive control methods which have been developed over the past twenty years and which use more than three parameters. Among adaptive control approaches, the pole placement (P.P.), the linear quadratic Gaussian (L.Q.G.), the generalized predictive adaptive control (G.P.A.C.) (Clarke et al., 1984) methods and learning systems seem to be the most attractive ones to provide automatic control parameters, hence minimizing the effect of variations in plant dynamics as well as disturbances. The first approaches are derived in the self-tuning manner (the unknown parameters are replaced by their estimates in the control signal synthesis) by solving a polynomial equation or by minimizing a quadratic criterion. In the second approach, a stochastic automaton and the random environment (the process to be controlled) are connected in a feed-back loop where the input to one is the output to the other (Najim, 1982). The P.P., L.Q.G., and the G.P.A.C self-tuning controllers, are composed of two loops. The inner loop consists of the process and an ordinary linear feed-back controller. The parameters of the model are updated by the other loop.

A learning system contains three loops: a simple feed-back

loop, an adaptive loop, and a learning loop. The schematic diagrams of these approaches are given in figure 1 and 2.

These control techniques have been successfully used in different fields such as the chemical and metallurgical industries, glasshouse heating systems, telephone traffic routines, resources allocation, etc.

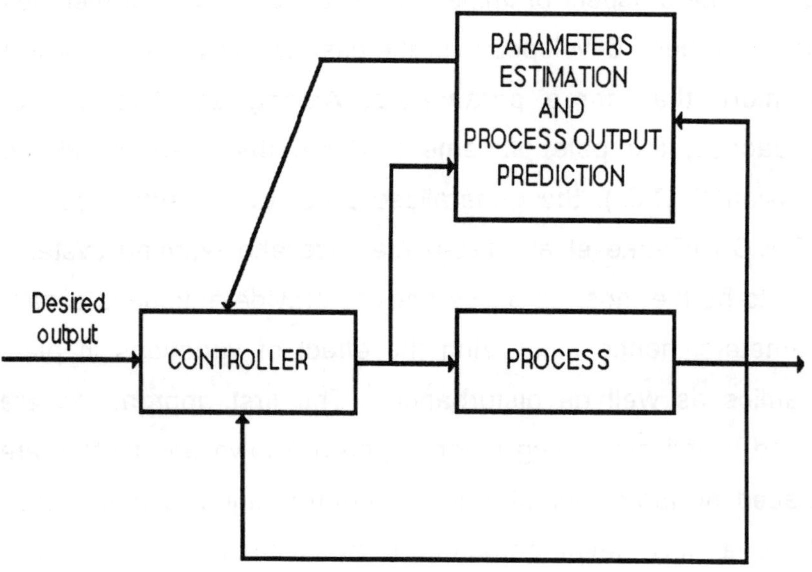

Figure 1. The adaptive controller

In this chapter we shall be concerned firstly with the pole placement, the linear quadratic Gaussian and the generalized predictive control algorithm which generalizes the minimum variance regulators and controllers (Åström and Wittenmark, 1973;

Clarke and Gawthrop, 1979) and secondly with the application of the learning automaton to process control.

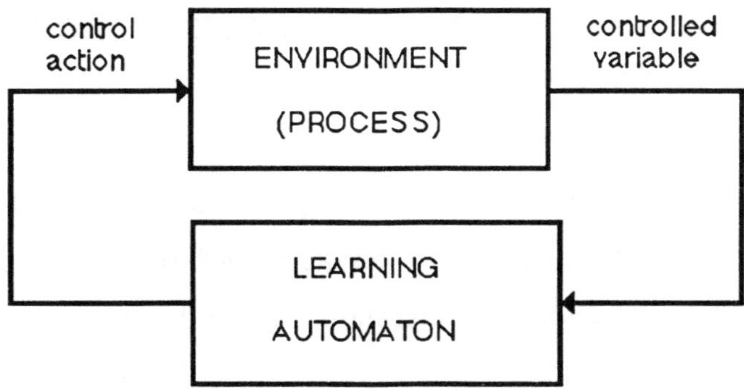

Figure 2. The learning automaton

The idea behind the pole placement technique is to derive a controller so that the closed loop system has a dynamics specified in advance. In the linear quadratic control approach the control action is derived from the minimization of a quadratic function to make a trade-off between the rate of decay of the state and the energy of the input. The basic idea used in the generalized predictive control algorithm is prediction. In fact if we shall focus our development on robust algorithms. Then the prediction problem will be detailed and solved and the generalized predictive adaptive control algorithm will be introduced. The last part of this chapter will deal with learning systems and their use for control purposes.

2. POLE PLACEMENT

2.1. INDIRECT ADAPTIVE POLE PLACEMENT CONTROL ALGORITHM

In spite of the fact that all practical processes are non-linear, good control performances can be achieved using white box models carried out by local linearization around the nominal or the considered operating point.

In this section we outline the development of a robust pole placement control algorithm. The pole placement technique consists of auto-placing the desired closed-loop poles. Let us consider a linear discrete model which is appropriate for digital control.

$$A(q^{-1})y(t) = q^{-1}B(q^{-1})u(t) \tag{1}$$

where $u(t)$ and $y(t)$ represent respectively the process input and output. k is the (integer) system time delay. $A(q^{-1})$ and $B(q^{-1})$ are coprime polynomials (no common zeros) in the backward shift operator q^{-1} $(q^{-i}y(t) = y(t - iT)$; (where T represents the sampling period) given by:

$$A(q^{-1}) = 1 + a_1 q^{-1} + \ldots + a_{na}q^{-na}$$

$$B(q^{-1}) = b_0 + b_1 q^{-1} + \ldots + b_{nb}q^{-nb} \tag{2}$$

Then, the transfer function relating the control action u(t) and the process output y(t) is

$$H(q^{-1}) = A(q^{-1})/B(q^{-1}) \qquad (3)$$

The dynamics of a process can be modified by feedback and depends on its poles. Many pole placement algorithms have been proposed and used successfully in industrial processes. We shall carry out a control strategy which moves the poles of the system to a desired domain to achieve good response such as fast rise time, minimum overshoot or quick settling time. To do this, let us consider the following control strategy:

$$R(q^{-1})\, u(t) = T(q^{-1})w(t) - S(q^{-1})\, y(t) \qquad (4)$$

where R, T and S are polynomials in q^{-1} of nr, nt and ns degrees.

The controller action u(t) is a linear function of the past values of the control signal, the output y(t), the desired output w(t) and their past values. This control action may place the closed-loop poles anywhere, for example within the unit circle. If the polynomials are correctly selected, this control law is able to control open-loop unstable nonminimum phase systems.

Figure 3 shows the block diagram of a closed loop pole placement controller.

Let us denote by $P(q^{-1})$ the desired closed loop characteristic polynomial (the zeros of the polynomial $P(q^{-1})$ are the desired poles of the closed loop system, i.e. the process and the regulator). The closed loop characteristic equation is given by:

$$A(q^{-1})R(q^{-1}) + q^{-k} S(q^{-1})B(q^{-1}) = P(q^{-1}) \qquad (5)$$

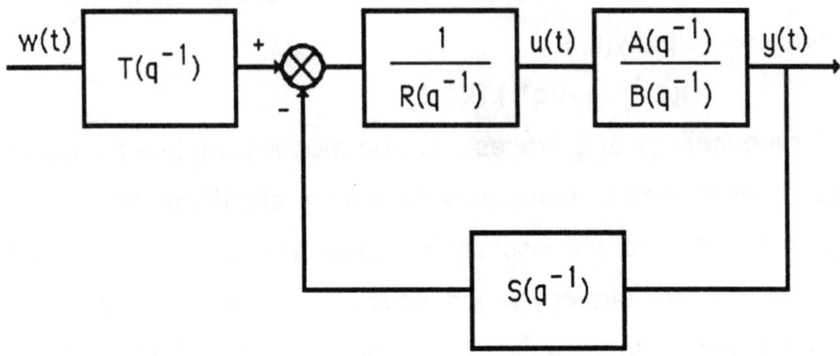

Figure 3. Block diagram of a pole placement regulator

where the polynomial $P(q^{-1})$ which specifies the desired closed loop pole locations is defined as:

$$P(q^{-1}) = 1 + p_1 q^{-1} + ... + p_{np} q^{-np} \qquad (6)$$

The design problem is equivalent to the algebraic problem of finding polynomial R and T such that equation (5) holds. Many methods (Kucera, 1979) have been proposed to solve the polynomial equation (5), (which is called a linear Diophantine equation).

A general solution of this equation can be obtained from any particular solution. The one which minimizes the degree of one polynomial (R or T) is called the minimum-degree solution with respect to the considered polynomial. The minimum solution with respect to R and with respect to T is generally not identical. The following conditions on the degrees ensures the existence and uniqueness of the Diophantine equation solutions:

1. If deg $P(q^{-1})$ < deg $A(q^{-1})$ + deg $B(q^{-1})$ + k - 1, two possible minimum degree solutions with respect to S and R respectively can be obtained:

The minimum degree solution with respect to S corresponds to:

$$\deg S(q^{-1}) = \deg A(q^{-1}) - 1$$
$$\deg R(q^{-1}) = \deg B(q^{-1}) + k - 1 \qquad (7)$$

2. If deg $P(q^{-1})$ > deg $A(q^{-1})$ + deg $B(q^{-1})$ + k - 1, two possible minimum degrre solutions with respect to S and R respectively can be performed:

The minimum degree solution with respect to S is:

$$\deg A(q^{-1}) = \deg P(q^{-1}) \text{ or } \deg R(q^{-1}) = \deg P(q^{-1}) - \deg A(q^{-1}) >$$
$$\deg B(q^{-1}) + k - 1$$
$$\deg R(q^{-1}) + \deg S(q^{-1}) + 1 = \deg P(q^{-1}) \quad \deg S(q^{-1}) =$$
$$\deg A(q^{-1}) + k - 1 \qquad (8)$$

If the minimum degree solution with respect to the degree of R is desired, the following degree conditions must be satisfied:

$$\deg R(q^{-1}) + \deg B(q^{-1}) + k = \deg P(q^{-1})$$
$$\text{or } \deg S(q^{-1}) = \deg P(q^{-1}) - \deg B(q^{-1}) > \deg A(q^{-1}) - 1 \qquad (9)$$
$$\deg R(q^{-1}) + \deg S(q^{-1}) + 1 = \deg P(q^{-1}) \quad \deg R(q^{-1}) =$$
$$\deg B(q^{-1}) + k - 1$$

Åström and Zhou (1981) have proposed an algorithm to solve the Diophantine equation in the cases of relative prime polynomials and a common factors existence. This algorithm uses the Euclidean algorithm (Kucera 1979) and a linear transformation.

<u>Example 1:</u>

Consider a process described by the polynomials

$$A(q^{-1}) = 1 + 1.55q^{-1} + 0.325q^{-2}$$

$$B(q^{-1}) = 1 + 0.4q^{-1}$$

The time delay and the desired characteristic equation are:

$$k = 1$$

$$P(q^{-1}) = 1 + q^{-1} + 0.21q^{-2}$$

The condition 1 is satisfied, then the minimum degree solution is:

$$R(q^{-1}) = 1 + 0.311q^{-1} \; ; \; S(q^{-1}) = -0.861 - 0.252q^{-1}$$

An obvious alternative solution to the Diophantine equation is to solve the algebraic system of equations derived from equating cefficients of the same power of the polynomial variable q^{-1} on both sides of equation (5). This system of linear equations, the solution to which can be performed using any classical method (Gauss elimination, factorization, etc.), may be written in the following compact matrix form:

$$\mathbf{A} X = Y \tag{10}$$

where

$$
A = \begin{bmatrix} 1 & & & & & & 0 \\ a_1 & & & & b_0 & \\ & & & 1 & & & 0 \\ & & a_1 & & & b_0 \\ & & & b_{nb} & & \\ a_{na} & & & & & \\ & & & & b_{nb} \\ & & a_{na} & & \\ 0 & 0 & & & \end{bmatrix}, \quad X = \begin{bmatrix} r_1 \\ \cdot \\ \cdot \\ r_{nr} \\ s_0 \\ \cdot \\ \cdot \\ s_{ns} \end{bmatrix}, \quad Y = \begin{bmatrix} p_1 - a_1 \\ p_2 - a_2 \\ \cdot \\ \cdot \\ p_{np} \\ 0 \\ \cdot \end{bmatrix}
$$

The matrix **A** is known as the Sylvester (or eliminant) matrix. According to Sylvester theorem (Wolovich 1974), if the polynomials A and B are relatively prime the matrix A is nonsingular.

An integral action must be introduced in the control law to ensure offset-free (to eliminate steady state error). Let us derive the steady state input-ouput relation. To do this it is sufficient to replace the operator q^{-1} in the closed loop transfer function by 1: ($x(t) = x(t - i)$ for all i, or, $x(t - i)$ is equal to $q^{-1}x(t)$, then q^{-1} may be set equal to 1). The closed loop transfer function is given by:

$$y(t) = q^{-k} B(q^{-1})T(q^{-1})w(t)/[A(q^{-1})R(q^{-1}) + q^{-k} S(q^{-1})B(q^{-1})] \quad (11)$$

To provide the offset-free performance property, it necessary to choose the control design polynomials such that $[B(1)T(1)/P(1)]$ are equal to unity.

When the model parameters are unknown or time-varying, recursive identification methods which concern the problem of computing models of dynamical processes, based on observed input-output data (measurements), can be implemented to estimate these parameters on-line. Then, the control design may be based on the estimates of these parameters. The model's unknown parameters may be replaced by their estimates to compute the polynomials R and S by solving the Diophantine equation. If the desired closed-loop characteristic equation is specified, the adaptive pole placement algorithm that achieves the goals described in this section can be described as follows:

Step 1: Using an identification procedure, estimate the coefficients of the polynomials $A(q^{-1})$ and $B(q^{-1})$ in the model (equation (1)).

Step 2: Determine the control law (equation (4)) parameters (coefficients of the polynomials $R(q^{-1})$ and $S(q^{-1})$) by solving the polynomial equation (5).

Step 3: Compute the control signal u(t) from equation (4).

Repeat steps 1-3 at each sampling time.

It should be noticed that indirect adaptive scheme are more useful for control of long duration processes, when the identification procedure can be supervized on-line to deal with problems such as parameter drift, non stabilizability, etc.

The estimation procedure can lead to polynomial $A(q^{-1})$ and $B(q^{-1})$ which have common zeros. To deal with this problem, we shall present a useful and robust pole placement algorirhm in the following.

2.2. REGULARIZED POLE PLACEMENT ADAPTIVE CONTROL ALGORITHM

Numerical instability can be the consequence of estimation of parameters or can result when recursive calculations are carrying out using computers where each arithmetic operation generally affected by roundoff error. These errors or approximations arise because the Central Processing Unit (CPU) machine can only represents a subset of real numbers. To ensure robustness in industrial applications, near singularity of the Diophantine equation must be avoided. In order to prevent pole-zero cancellation and consequently to be able to solve the control equation design (equation (5), it is necessary to develop a procedure to obtain a regularized solution to the control problem.

The solution of this problem can be performed by applying the perturbation theory (Bauer-Fike theorem, etc.) or by drawing on the studies related to linear system equations sensitivity analysis (Golub and Van Loan, 1983).

One solution of this ill-conditioned problem can be obtained (Youlal et al., 1988) by doping, in a way, the Sylvester matrix. This idea is closely connected with the robust pole placement algorithm which consists of solving the following matrix equation instead of the classic polynomial equation (5):

$$(\mathbf{A} + \epsilon\mathbf{I})X = Y \tag{12}$$

where $\mathbf{I}$ is the identity matrix and the small positive number ϵ is named the regularization parameter.

If the polynomials $A(q^{-1})$ and $B(q^{-1})$ are common factors, the determinant of the matrix $\mathbf{A}$ vanishes. On the other hand, the matrix is nonsingular. The parameter can be chosen so as to keep the Sylvester matrix at a certain distance from singularity and to ensure boundedness of the controller parameters. The following simple rule of thumb, which has often been tested can be used:

$$\epsilon = \propto \| \mathbf{A} \| \tag{13}$$

In the above equation $\| \ \|$ denotes matrix norm. å is a positive scalar which can be chosen in the interval $[\,10^{-4}, 10^{-2}\,]$.

To deal with common zero and poles, another approach can be performed. It consists of performing the correction on the estimated polynomials $A(q^{-1})$ or $B(q^{-1})$ instead of on the matrix equation (10) (Youlal et al., 1988). This method is summarized as follows:

- Check if a common factor to the polynomials $A(q^{-1})$ and $B(q^{-1})$ exists, by either monitoring the determinant of the Sylvester matrix or by using the Euclidean algorithm (Kucera 1979).

- Regularize the control design equation (5) by adding a correction factor to the highest degree coefficient of either polynomial $A(q^{-1})$ or $B(q^{-1})$, i.e., make for instance

$$A(q^{-1}) = 1 + a_1 q^{-1} + \ldots + (a_{na} + \beta)\, q^{-na} \tag{14}$$

β is the regularization factor. It is a positive small scalar which can be chosen according the following rule of thumb:

$$\beta = \epsilon \, \| A(q^{-1}) \|_p \tag{15}$$

$\| \cdot \|_p$ denotes any suitable polynomial norm, and is some small positive number to be chosen so as to remove the ill-conditioning of the control design equation.

The regularized pole placement adaptive control algorithm and a recursive identification algorithm may be readily combined into a self-tuning algorithm as follows:

Data: Choose the initial settings of the estimator, the desired closed-loop characteristic polynomial, the regularization parameter and the threshold of the determinant to detect singularity or near singularity (determinant close to zero) of the Sylvester matrix.

Step 1: Estimate the model parameters (using a recursive method from the previous chapter).

Step 2: build the Sylvester matrix and check if regularization is needed.

Step 3: Calculate controller polynomials by solving equation (5) or its equivalent (matrix equation (10)) based on the parameter estimates.

Step 4: Calculate and implement new control signal u(t) (equation (4)).

Step 5: Save the data for the next sampling period.

Step 6: Go to step 1 at next sample instant.

Notice that the presented algorithm reduces to a standard adaptive pole placement scheme when the polynomial $A(q^{-1})$ and $B(q^{-1})$ have no common zeros (well-conditioned equation (5)). We have devoted more attention to the regularization procedures than to the pole placement algorithm itself, because the ill-conditioning problem of the control design equation is the most important point in adaptive pole placement algorithms.

The presented regularization procedure is very simple and can be combined with any indirect adaptive pole placement control approach to enhance its performances, particularly during the transient phase (start-up, set point variation, etc.) when the parameter estimates are still uncertain. This procedure does not lead to an increased complexity of computation.

We shall be concerned in the following section with the development of a Linear Quadratic Gaussian (L.Q.G.) controller with restricted stability domain. The restricted stability domain is a subset of the unit circle (in the discrete case) or a subset of the half left plane (in the continuous case).

3. LINEAR QUADRATIC GAUSSIAN CONTROLLER WITH RESTRICTED STABILITY DOMAIN

Many adaptive control algorithms have been proposed and used in industrial contexts. The L.Q.G. approach ensures stability of the closed loop system when the process parameters under consideration are known. This property is useful because generally, the models obtained are nonminimum phase, essentially owing to discretization.

Nevertheless, the closed loop dynamics are generally not controlled by the L.Q.G.. A controller system using a pure L.Q.G. does not provide optimal performance under all operating conditions because the quadratic criterion behind the L.Q.G. control strategy is the best compromise between conflicting requirements. The choice of the weighting factor may be difficult. When the process has no unstable zero, this factor can be set to zero. Lam (1982) proposes a method for rescaling the weighting factor in order to avoid gain variation (the appropriate choice of the weighting factor depends on the zero localization) and consequently not to make the closed loop poles move from their positions. He has suggested that one should adapt this factor as follows:

$$\lambda = \lambda_0 \, [B(1)]^2 \tag{16}$$

The choice of the parameter λ_0 in the domain (0, 1) leads to good performances (Lam, 1982).

In the Amerongen et al., (1986) method, the control requirements are translated into the weighting factor.

In this section we investigate a novel approach, which ensures that the closed loop poles lie in a domain included in the unit circle (stability domain) and which is defined a priori. The system under consideration is single-input, single-output and is described by the model:

$$A(q^{-1})y(t) = q^{-1}B(q^{-1})u(t) + e(t) \tag{17}$$

where $u(t)$, $y(t)$ and $e(t)$ denote the input, the output and a bounded and unmeasured perturbation.

Several state space representations exist to describe internal and external behavior of linear finite-dimensional systems. The representation where the time delay appears explicitly will be adopted in the following.

$$x(t+1) = A_s x(t) + B_s u(t-k+1) + D_s e(t)$$

$$y(t) = C_s^T x(t) \tag{18}$$

where the subscript s denotes the state space representation, A_s is a matrix, and B_s, D_s, C_s and x are vectors. These vectors and matrix are defined as follows:

$$
A_S = \begin{bmatrix}
-a_1 & -a_2 & \cdots & -a_{na} & b_1 & \cdots & b_{nb} \\
1 & 0 & \cdots & 0 & \cdots & & 0 \\
0 & 1 & 0 & & & & 0 \\
& & \cdot & & \cdot & & \\
& & \cdot & & \cdot & & 1 \\
0 & & & & 0 & \cdot & 0 \\
0 & & & & 1 & \cdot & 0 \\
& & \cdot & & \cdot & & \\
0 & 0 & \cdot & & & 1 & 0
\end{bmatrix}, \quad
B_S = \begin{bmatrix}
b_0 \\
0 \\
\cdot \\
0 \\
1 \\
0 \\
\cdot \\
0
\end{bmatrix}
$$

$$x(t) = [y(t)\, y(t-1), \ldots, y(t - na + 1)\, u(t-k), \ldots, u(t - k - nb + 1)]^T$$

$$C_S = [\,1, 0, \ldots, 0\,]^T \quad , \quad D_S = [\,1, 0, \ldots, 0\,]^T$$

Notice that the state vector components are all measurable.

In practical applications it is useful to consider a relationship between the input and the output variations as a model for control purposes. These variations and the corresponding model may be defined as follows:

$$A_v(q^{-1})y_v(t) = q^{-k} B_v(q^{-1})\, u_v(t) + b(t) \tag{19}$$

with

$$y_v(t) = y(t) - w(t)$$

$$u_v(t) = \Delta(q^{-1})\, u(t) \tag{20}$$

where

$\Delta(q^{-1}) = 1 - q^{-1}$ is the differencing operator which permits the introduction of integral action in the control law ensure that it is offset-free. The subscript v denotes the variation and b(t) represents the disturbance affecting the process behavior. This representation is called a performance model by M'Saad (1987).

This implies, using equations (17) and (19 - 20), that

$$A_V(q^{-1}) = \Delta(q^{-1}) A(q^{-1})$$

$$B_V(q^{-1}) = B(q^{-1}) \tag{21}$$

The corresponding parametrization as a stochastic dynamical system in state-space coordinates is given by :

$$x_V(t+1) = A_{sv}x_V(t) + B_s u_V(t-k+1) + D_s b(t)$$

$$y(t) = C_s^T x_V(t) \tag{22}$$

where

$$x_V(t) = [\, y_V(t), \ldots, y_V(t - na - 1), u_V(t - k), \ldots, u_V(t - k - nb + 1)\,]^T$$

Using the coefficients of the polynomials $A_V(q^{-1})$ and $B_V(q^{-1})$, the matrix A_{sv} can be generated as the matrix A_s.

Among other objectives (which will be specified in the following) , the purpose of the control is to minimize the loss function

$$J = \sum_{j=1}^{Nr(t)} \{y_V(t + j)^2 + \lambda\, u_V(t + j - 1)^2\} \tag{23}$$

where λ is the weighting factor on control and $Nr(t)$ is the iterations number associated with the Riccati equation. It will be defined in the sequel.

Another objective may be added to that represented by the

criterion (23). It consist of the requirements such as fast rise time, minimum overshoot or quick settling time. This desired closed loop behavior may be translated in terms of stability as follows:

To design a controller such that the closed loop poles lie in a certain domain (figure 4) included in the classic stability domain (the unit circle) will ensure better transient response.

This restricted stability is a circle of radius δ and centered at the point of coordinate ω on the real axis.

Let us note this domain by $C(\delta, \omega)$.

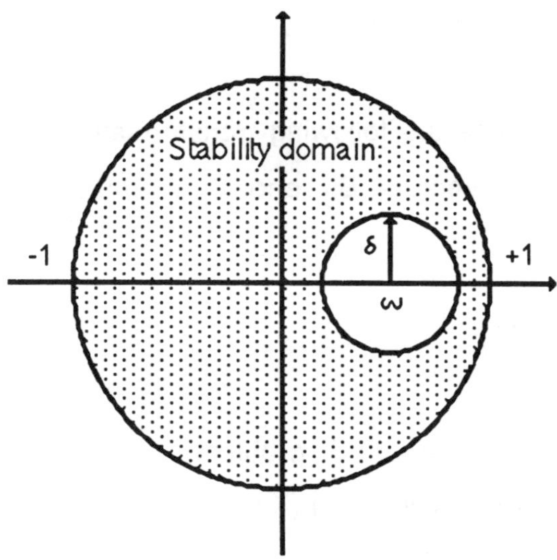

Figure 4. Restricted stability domain

Notice that a procedure for carrying out control laws that

give closed loop system with all poles inside a circle of radius δ smaller than unity and centered at the origin have been devised by Åström and Wittenmark (1984). This control strategy is based on the following criterion minimization:

$$J = E\{(1/\delta)^{2t}[y^2(k) + \lambda u^2(t)]\} \tag{24}$$

Using the control law which minimizes the criterion (24), the input and the output converge to zero at least as fast as r^t when t is increased (Åström and Wittenmark, 1984).

We shall develop in the following a solution of the L.Q.G. problem with restricted domain stability domain. The resulting algorithm generalizes the approach derived from the criterion (24) minimization.

Let us introduce the following transformations of the matrix A_{sv} and of the vectors B_s and C_s involved in the state representation given by equation (22):

$$A_r = \delta [A_{sv} - \omega I]$$

$$B_r = \delta B_s \tag{25}$$

The solution of the L.Q.G. problem, thus formulated is well known and can be simply stated (Kwakernaak, 1972; Franklin and Powell, 1980; etc.) as follows:

The optimal control u(t) which minimizes the quadratic performance index (23) can be expressed as a linear function of the state variable, $x_v(t)$; i.e.,

$$u_v(t) = - L (Nr(t)) x_v(t) \tag{26}$$

where L (Nr(t)) is the feedback gain given by:

$$L\ (Nr(t)) = [\ \lambda + B_r^T R(Nr(t) - 1)B_r\ ]^{-1}\ B_r^T R(Nr(t) - 1)\ A_r \quad (27)$$

$$R(Nr(t)) = A_r^T R^\circ(Nr(t))A_r\ C_s C_s^T \quad (28)$$

$$R^\circ(Nr(t)) = R(Nr(t) - 1) - R(Nr(t) - 1)B_r\ B_r^T R(Nr(t) - 1)[1 +$$

$$+ B_r^T R(Nr(t) - 1)B_r] \quad (29)$$

With the initial condition

$$R(0) = C_s C_s^T \quad (30)$$

$R(Nr(t))$ is a square matrix of dimension $(na + nb + 1).(na + nb + 1)$

The control law (equation (26)) is simply a feedback of a linear combination of all the state variables.

Equation (28) is called the discrete Riccati equation which also appears in the solution of estimation problems (Kalman filter). The controller system is represented pictoriallyin figure in figure 5, the shaded region denoting the original, open loop or uncompensated system. The derivation of the Linear Quadratic Controller is given in appendix A.

If the control law (equation (26)) is connected to the process to be controlled, the closed loop system is described by the state discrete equation (the disturbance is taked into consideration because the state does not need to be reconstructed) which can be found by substituting the control u(t) as given by equation (26) in equation (22).

$$x_v(t + 1) = [\ A_r - B_r\ L\ (Nr(t))\ ]\ x_v(t) \quad (31)$$

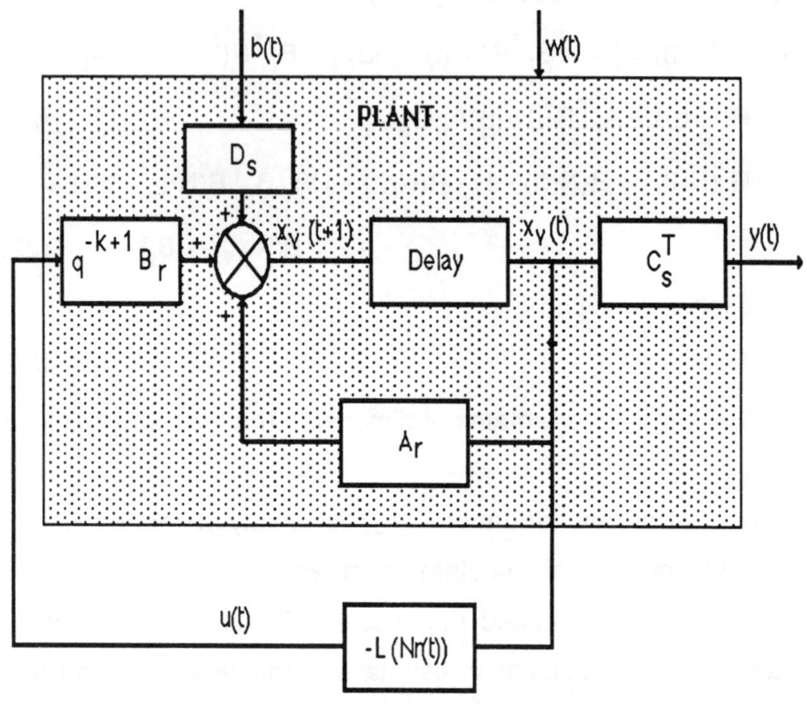

Figure 5. The L.Q.G. controller system

The parameter Nr(t) is increased from one until the eigenvalues associated to the closed loop system are less than δ in absolute value. These eigenvalues are any values λ_e such that the determinant of the matrix is; i.e. the zeros of $|\lambda_e I - A_r + B_r L(Nr(t)) + \omega|$ define the eigenvalues of the matrix $[A_r - B_r L(Nr(t)) - \omega]$. This means that when this condition is

76525101-89

fulfilled, the closed loop poles lie in the restricted stability domain.

The vector L (Nr(t)) is of dimension na + nb + 1; it may be noted as:

$$L\ (Nr(t)) = [l_1, l_2, \ldots, l_{na+nb+1}]^T$$

In order to perfom a stability test which indicates if the the desired closed loop dynamic requirements are achieved, i.e. ,the closed loop poles lie in the circle $C(\delta, \omega)$, we shall present the L.Q.G. controller as the solution of a pole placement problem where the purpose is to arrange the state feedback so that the all the closed loop system poles belong to the desired domain.

Rearrangement of equation (26) as equation (4) results in

$$R(q^{-1})u_v(t) = - S(q^{-1})\, x_v(t) \tag{32}$$

Thus the characteristic equation of the controlled (closed loop) system is

$$A(q^{-1})R(q^{-1}) + q^{-k} B(q^{-1})S(q^{-1}) = 0 \tag{33}$$

To be able to use any stability test (Kucera, 1979), we shall replace the the backward shift operator q^{-1} by $(\omega + \delta q^{-1})$.

<u>Example 2:</u>

Consider a polynomial $A(q^{-1})$ which has a zero equal to: $q^{-1} = 0.5 + 0.6\, j$ where j is the complex number $(j^2 = -1)$. The zero of this polynomial lies in the unit circle and not in the domain C(0.5, 0.5) which can be easily shown by performing the transformation $(\omega + \delta q^{-1})$ on the operator q^{-1}. This variable change leads to the the following polynomial $A^*(q^{-1}) = 1.2\,j + 0.5q^{-1}$ whose zero module is great than unity. This implies instability in the the sense defined above.

The cost function associated with the control law (equation (26)) is equal to

$$J\,(Nr(t))= x_V(t)^T \mathbf{R}(Nr(t))x_V(t)$$

The convergence analysis of the L.Q.G. controller shows that the closed loop system is stable if and only if the following condition is verified :

$$\lim_{Nr(t)\,\rightarrow\,\infty} J\,(Nr(t)) < \infty \ \text{ and } \ \|x_V(t)\| < \infty \tag{34}$$

The above condition is satisfied if

$$\text{trace } \mathbf{R}(Nr(t)) < \infty \tag{35}$$

Hence the trace of the matrix $\mathbf{R}(nr(t))$ can be used as indicator for on-line test of the stabilizabilty of the system. This problem is related to the appearance of an unstable zero equal to an unstable pole (simplification of unstable zero and poles).

4. MINIMUM MEAN SQUARE ERROR PREDICTION

In this section, the linear minimum mean square error prediction of a linear system driven by a random excitation will be presented.

Consider a linear stochastic system excited by a perturbation and described by the following equation:

$$A(q^{-1})y(t) = C(q^{-1})e(t) \qquad (36)$$

where

$$A(q^{-1}) = 1 + a_1 q^{-1} + \ldots + a_n q^{-n}$$

$$C(q^{-1}) = 1 + c_1 q^{-1} + \ldots + c_n q^{-n} \qquad (37)$$

$\{e(t)\}$ is a zero mean sequence of independent random variables with finite variance supposed to be equal to 1.

The problem to be solved is: at time t, taking into consideration the available data I (all the measurements), find a prediction $y^*(t+k/t)$ of $y(t+k)$ so that the mean square error is minimized.

$$J = E\{e_p(t+k)^2\} \qquad (38)$$

where $e_p(t+k)$ is the k-step prediction error

$$e_p(t+k) = y(t+k) - y^*(t+k/t) \qquad (39)$$

The parameter k represents the prediction horizon. The prediction $y^*(t+k/t)$ will be necessarily a function of the information I.

Now consider

$$y(t+k) = \{C(q^{-1})/A(q^{-1})\}e(t+k) \qquad (40)$$

The right-hand side term in equation (5) is a linear arrangement of $e(t+k), e(t+K-1), \ldots, e(t+1); e(t), e(t-1), \ldots$ The term $e(t)$ and its past can be calculated from the information I.

$$e(t) = \{A(q^{-1})/C(q^{-1})\}y(t) \qquad (41)$$

To separate the available data from the future measurements, let us introduce the following identity (Åström, 1970):

$$C(q^{-1})/A(q^{-1}) = F(q^{-1}) + q^{-k}G(q^{-1})/A(q^{-1}) \qquad (42)$$

where

$$F(q^{-1}) = 1 + f_1 q^{-1} + \ldots + f_{k-1}q^{1-k}$$

$$G(q^{-1}) = g_0 + g_1 q^{-1} + \ldots + g_n q^{1-n} \qquad (43)$$

The polynomials $F(q^{-1})$ and $G(q^{-1})$ are respectively the quotient and the remainder of the division $C(q^{-1})/A(q^{-1})$.

Equation (5) can then be written as

$$y(t+k) = G(q^{-1})/A(q^{-1})e(t) + F(q^{-1})e(t+k) \qquad (44)$$

It follows from (6) that

$$y(t+k) = \{G(q^{-1})/C(q^{-1})\}y(t) + F(q^{-1})e(t+k) \qquad (45)$$

The criterion J may be written as

$$J = \mathbf{E}\{[G(q^{-1})/C(q^{-1})]y(t) + F(q^{-1})e(t+k) - y^*(t+k/t)\} \qquad (46)$$

The cost prediction function will be minimized if we choose the prediction $y^*(t+k/t)$ equal to

$$y^*(t+k/t) = [G(q^{-1})/C(q^{-1})]y(t) \qquad (47)$$

The transfer function $[G(q^{-1})/C(q^{-1})]$, related to the output and its prediction must be stable. Then the zeros of the polynomial

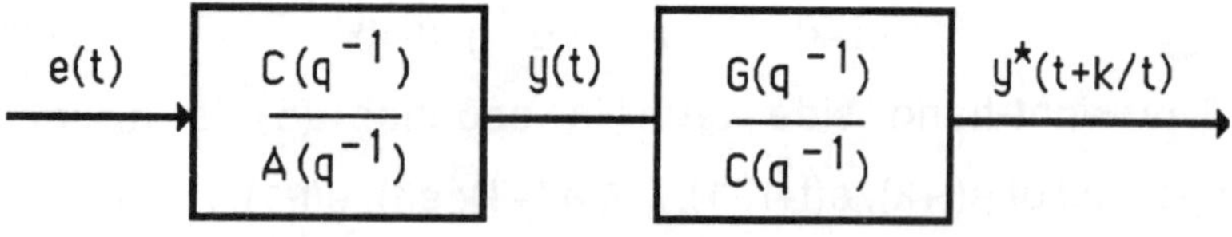

Figure 6. System and its output prediction

$C(q^{-1})$ must lie outside the unit circle. The schematic diagram of the system and its output prediction is given in figure 6.

The optimal prediction error is

$$e_p(t+k) = F(q^{-1})e(t + k) \qquad (48)$$

Then the minimum value of the criterion J is given by

$$J = 1 + \sum_{i=1}^{k-1} f_i^2 \qquad (49)$$

Note that the variance of the prediction error increases with the prediction horizon.

The identity (7) leads to a set of linear algebraic equations.

$$1 = 1$$

$$c_1 = a_1 + f_1$$

$$c_2 = a_2 + a_1 f_1 + f_2$$

$$\cdots \cdots \cdots$$

$$c_{k-1} = a_{k-1} + a_{k-2} f_1 + \ldots + a_1 f_{k-2} + f_{k-1} \qquad (50)$$

$$c_k = a_k + a_{k-1} f_1 + \ldots + a_1 f_{k-1} + g_0$$

$$\cdots \cdots \cdots$$

$$c_n = a_n + a_{n-1} f_1 + \ldots + a_{n-k+1} f_{k-1} + g_{n-k}$$

$$0 = a_n f_1 + \ldots + a_{n-k} f_{k-1} + g_{n-k+1}$$

$$\cdots \cdots \cdots$$

$$0 = a_n f_{k-1} + g_{n-1}$$

Let us write equation (15) in a matrix form

$$
\begin{bmatrix}
1 & & & & & \\
a_1 & 1 & & & & \\
a_2 & a_1 & 1 & & & \\
\cdot & \cdot & \cdot & \cdot & & \\
a_k & a_{k-1} & & 1 & & \\
\cdot & \cdot & & \cdot & & \\
a_n & a_{n-1} & \cdot & \cdot & \cdot & 1 \\
\cdot & \cdot & \cdot & & & \\
\cdot & \cdot & \cdot & & \cdot & \\
& & & & & \\
\cdot & \cdot & \cdot & & & 1
\end{bmatrix}
g_0 =
\begin{bmatrix}
1 \\
f_1 \\
f_2 \\
\cdot \\
g_0 \\
\cdot \\
\cdot \\
\cdot \\
\cdot \\
\\
g_n
\end{bmatrix}
=
\begin{bmatrix}
1 \\
c_1 \\
c_2 \\
\cdot \\
c_k \\
\cdot \\
c_n \\
\cdot \\
\cdot \\
\\
0
\end{bmatrix}
\tag{51}
$$

The above triangular matrix has an inverse, so the solution of equation (16) exists and is unique.

Example 3:

Given the system

$$y(t) + 1.1y(t - 1) + 0.28y(t - 2) = e(t) + 0.4e(t - 1) - 0.05e(t - 2)$$

Determine the k-steps predictor of the output $y(t)$ for $k = 2$ and 3.

The solution of the identity (6) gives the following:

For k=2 we obtain:

$$F(q^{-1}) = 1 - 1.1q^{-1} \quad ; \quad G(q^{-1}) = 0.44 + 0.196q^{-1}$$

$$y^*(t + 2/t) + 0.4y^*(t + 1/t - 1) - 0.05y^*(t/t - 2) = 0.44y(t) + 0.196y(t - 1)$$

The variance of the error prediction is $J = 1.49$

<u>For k=3 we obtain:</u>

$$F(q^{-1}) = 1 - 0.7q^{-1} + 0.44q^{-2} \; ; G(q^{-1}) = 0.288 + 0.1232q^{-1}$$

$$y^*(t + 2/t) + 0.4y^*(t + 1/t - 1) - 0.05y^*(t/t - 2) = 0.288y(t) + 0.1232y(t - 1)$$

The variance of the error prediction is $J = 1.6836$ which is great than this, corresponding to a time delay equal to 2.

5. CONTROL OBJECTIVE

Figures 7 shows a classic recorder process output and figure 8 shows the evolution of output characteristics (moisture content, impurity, etc.) of the process as a function of the energy consumption. This variation has an exponential profile, and becomes flat for a weak impurity degree.

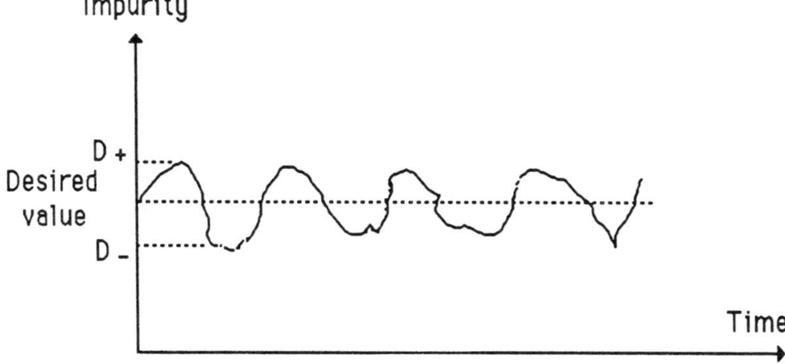

Figure 7. Time evolution of the impurity degree

We can see that the mean value of the output is constant, but, at the same time, when the product quality is less than the desired value, the energy saving does not compensate the energy which is left over. The energy consumption is consequently not optimized. It is also necessary to include in the cost function two

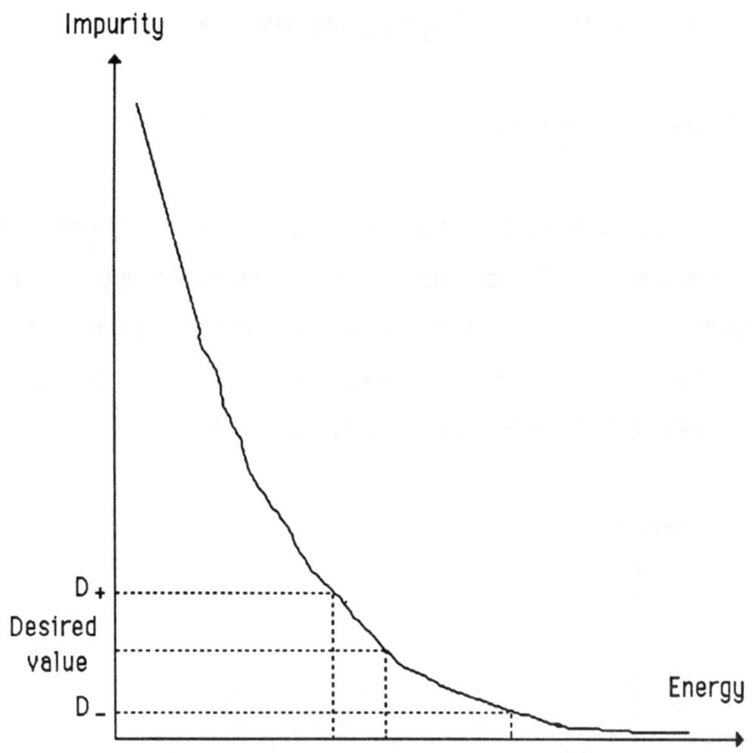

Figure 8. Evolution of the impurity degree as a function of the energy

other terms, one related to the fluctuation of the output near the desired value and the other related to the energy. Taking these objectives into consideration, we obtain a control criterion which has a strong intuitive appeal to the operator, is quadratic in deviation of the process output from its set point and also quadratic in control.

6. GENERALIZED PREDICTIVE CONTROL

Many self-tuning algorithms using process output prediction have been proposed over the last decade and have been applied in different industries (cement manufacture, phosphate drying, paper manufacture, chemistry, biotechnology, etc.). In several self-tuning prediction and control methods, the prediction is carried out over a single stage requiring the knowledge of the time delay. As the time delay is not commonly well known and is possibly time-varying, it is necessary at each time to make a forecast of the plant output over a long range horizon (Ydstie, 1982); i.e., the prediction horizon should be greater than the equivalent time delay (the actual time delay augmented by the unstable zeros number). The generalized predictive control algorithm (G.P.C.) developed by Clarke et al., (1984) provides the most general structure of the long range predictive control methods. That is why we will present such a design method in the following.

6.1. PROCESS MODEL AND CONTROL OBJECTIVE

A discrete-time model of the process and disturbance dynamics is given by the following equation:

$$A(q^{-1})y(t) = q^{-k}B(q^{-1})u(t) + e(t)/\Delta(q^{-1}) \qquad (52)$$

With $A(q^{-1}) = 1 + a_1 q^{-1} + \ldots + a_n q^{-n}$

$$B(q^{-1}) = b_0 + b_1 q^{-1} + \ldots + b_n q^{-n}$$

$$\Delta(q^{-1}) = 1 - q^{-1}$$

and $y(.)$ and $u(.)$ process output and input, k the time delay (delay=k.T, where T is the sampling period) and $\{e(t)\}$ an uncorrelated stochastic process with zero mean and finite variance.

The disturbance model $\{e(t)/\Delta(q^{-1})\}$ is more suitable for industrial process disturbance representation (Tuffs and Clarke, 1985). In the steady state, the process output should be equal to the desired value or setpoint. An appropriate solution to achieve this property consists of incorporating integral action in the control law. Such a design feature is ensured by the $\Delta(q^{-1})$ operator.

The design problem is to find the control which gives good performance of the closed loop system. A good trade-off between energy and output behaviour (to be close to the desired value) is often achieved if the control objective is chosen in terms of minimizing the cost function:

$$J = \mathbf{E}\left[\left\{\sum_{j=k}^{Ny} [(y(t+j) - w(t+j))^2 + \lambda\ \Delta u^2(t+j-k)]\right\}/t\right] \qquad (53)$$

where Ny is the control horizon

    w(.) the desired output sequence

    λ the control-weighting factor

    E mathematical expectation operator

To derive the optimal control in the sense of criterion (18), let us first write the output at time $t + j$.

$$y(t+j) = [q^{-k}B(q^{-1})/A(q^{-1})]u(t+j) + [1/\Delta(q^{-1})A(q^{-1})]e(t+j) \qquad (54)$$

The second term in the right hand side of equation (19) is a function of $\{e(t+j)$,

$e(t+j-1),...., e(t+1); e(t), e(t-1),...\}$. The quantity $\{e(t), e(t-1),.....\}$ can be computed from the available data at time t. The available data can be manipulated as we wish. We shall introduce an identity to make a separation between these two kinds of data and to express equation (19) in an equivalent prediction model form.

$$1 = E_j(q^{-1})A(q^{-1})\Delta(q^{-1}) + q^{-j}F_j(q^{-1}) \qquad (55$$

where $E_j(q^{-1}) = 1 + e_{j,1}q^{-1} + ... + e_{j,j-1}q^{-j+1}$

    $F_j(q^{-1}) = f_{j,0} + f_{j,1}q^{-1} + ... + f_{j,n}q^{-n}$

Now it is possible to rewrite the system equation in a prediction form. Multiplying equation (19) by $\{E_j(q^{-1})A(q^{-1})\Delta(q^{-1})\}$, it follows that:

$$E_j(q^{-1})A(q^{-1})\Delta(q^{-1})y(t+j) = q^{-k}E_j(q^{-1})B(q^{-1})\Delta(q^{-1})u(t+j) +$$
$$+ E_j(q^{-1})e(t+j) \qquad (56)$$

Using identity (20) gives:

$$y(t + j) = q^{-k}EB_j(q^{-1})\Delta(q^{-1})u(t + j) + F_j(q^{-1})y(t) + E_j(q^{-1})e(t + j) \qquad (57)$$

where $EB_j(q^{-1}) = E_j(q^{-1})B(q^{-1})$

Let write equation (22) for $j = k$ to Ny

$$y(t + k) = EB_k(q^{-1})\Delta(q^{-1})u(t) + F_k(q^{-1})y(t) + E_1(q^{-1})e(t + k)$$

$$y(t + k + 1)=EB_{k+1}(q^{-1})\Delta(q^{-1})u(t + 1)+F_{k+1}(q^{-1})y(t)+$$

$$+ E_{k+1}(q^{-1})e(t + k + 1)$$

$$\ldots\ldots\ldots$$

$$\qquad (58)$$

$$y(t + Ny) = EB_{Ny}(q^{-1})\Delta(q^{-1})u(t + Ny - k) + F_{Ny}(q^{-1})y(t) +$$

$$+ E_{Ny}(q^{-1})e(t + Ny)$$

We shall divide the terms $EB_j(q^{-1})\Delta(q^{-1})u(t + j - k)$ into past and future components, then equation (23) may be written in a matrix form

$$Y = G\Delta U + F + E \qquad (59)$$

where

$$Y = [y(t + k),\ldots., y(t + Ny)]^T$$

$$E = [E_k(q^{-1})e(t + k),\ldots., E_{Ny}(q^{-1})e(t +Ny)]^T$$

$$F = [F_k(q^{-1})y(t) + \text{the known terms of } EB_k(q^{-1})u(t),\ldots.]^T$$

$$\Delta U = [\Delta u(t),\Delta u(t + 1),\ldots.,\Delta u(t + Ny - k)]^T$$

G is a lower triangular matrix.

Example 4:

Consider a system given by (17), where

$$A(q^{-1}) = 1 + 0.9q^{-1} + 0.14q^{-2}$$

$$B(q^{-1}) = 1 + 0.5q^{-1}$$

$$k=1$$

For $j = 1$ to 4 ($Ny = 4$), we obtain the following expressions for $E_j(q^{-1})$, $EB_j(q^{-1})$, $F_j(q^{-1})$ and for The matrix G and the vector F.

$$E_1(q^{-1}) = 1$$

$$F_1(q^{-1}) = 0.1 + 0.76q^{-1} + 0.14q^{-2}$$

$$E_2(q^{-1}) = 1 + 0.1q^{-1}$$

$$F_2(q^{-1}) = 0.77 + 0.216q^{-1} + 0.014q^{-2}$$

$$E_3(q^{-1}) = 1 + 0.1q^{-1} + 0.77q^{-2}$$

$$F_3(q^{-1}) = 0.293 + 0.5992q^{-1} + 0.1078q^{-2}$$

$$E_4(q^{-1}) = 1 + 0.1q^{-1} + 0.77q^{-2} + 0.293q^{-3}$$

$$F_4(q^{-1}) = 0.628 + 0.3304q^{-1} + 0.041q^{-2}$$

$$EB_1(q^{-1})\Delta(q^{-1})u(t) = \Delta u(t) + \{0.5\Delta u(t-1)\}$$

$$EB_2(q^{-1})\Delta(q^{-1})u(t+1) = \Delta u(t+1) + 0.6\Delta u(t) + \{0.05\Delta u(t-1)\}$$

$$EB_3(q-1)\Delta(q^{-1})u(t+2) = \Delta u(t+2) + 0.6\Delta u(t+1) + 0.82\Delta u(t) +$$

$$+ \{0.385\Delta u(t-1)\}$$

$$EB_4(q-1)\Delta(q^{-1})u(t+3) = \Delta u(t+3)+0.6\Delta u(t+2) + 0.82\Delta u(t+1) +$$

$$+ 0.678\Delta u(t) + \{0.1465\Delta u(t-1)\}$$

The corresponding matrix G and vector ΔU are given by

$$G = \begin{bmatrix} 1 & 0 & 0 & 0 \\ 0.6 & 1 & 0 & 0 \\ 0.82 & 0.6 & 1 & 0 \\ 0.678 & 0.82 & 0.6 & 1 \end{bmatrix} \quad \Delta U = \begin{bmatrix} \Delta u(t) \\ \Delta u(t+1) \\ \Delta u(t+2) \\ \Delta u(t+3) \end{bmatrix}$$

A recursion algorithm for the identity (20) resolution developped by Clarke and co-workers is given in the appendix B. It is realized by the Fortran subroutine RIDEN given below.

```
      SUBROUTINE RIDEN(A,AB,E,F,N1,J)
c
c     RESOLUTION OF THE IDENTITY (EQUATION (20))
c
c     N    THE DEGREE OF THE POLYNOMIAL A(q-1)
c     N1   THE COEFFICIENTS OF THE POLYNOMIAL A(q-1)
c          (N1=N+1)
c     A    VECTOR OF DIMENSION N1 CONTAINING THE
c          COEFFICIENTS OF THE POLYNOMIAL A(q-1)
c     AB   VECTOR OF DIMENSION (N1+1) RELATED TO THE
c          POLYNOMIAL A(q-1)A(q-1)
c     E    VECTOR OF DIMENSION J CONTAINING THE
c          COEFFICIENTS OF THE  POLYNOMIAL Ej(q-1)
c     F    VECTOR OF DIMENSION (N+1) CONTAINING THE
```

```fortran
c          COEFFICIENTS OF THE POLYNOMIAL Fj(q-1)
c          THE POLYNOMIAL E AND F MUST BE INITIALIZED IN
c          THE HOST PROGRAM
c
      DIMENSION A(N1),AB(N1+1),E(J),F(N1)
c
      AB(1)=1.0
      AB(N1+1)=-A(N1)
      DO 10 i=2,N1
   10 AB(i)=A(i)-A(i-1)
      IF(J.GT.1) GOTO 11
      E(1)=1.0
      DO 12 i=1,N1
   12 F(i)=-AB(i+1)
      GOTO 13
   11 CONTINUE
      E(J)=F(1)
      DO 14 i=1,N1
   14 F(i)=F(i+1)-AB(i+1)*E(J)
   13 CONTINUE
      RETURN
      END
```

We shall now look at the development of the optimal control strategy.

According to equation (24), the criterion J may be written as

$$J = E\{[G\Delta U + F + E - W]^T[G\Delta U + F + E - W] + \lambda\Delta U^T\Delta U\} \qquad (60)$$

where

$$W = [w(t + 1), w(t + 2),....,w(t + Ny)]^T$$

The disturbance E is independent of the available data and by assumption uncorrelated with the desired values W. The minimization of the cost function J leads to the optimization of the following deterministic function:

$$J_d = \{[G\Delta U + F - W]^T[G\Delta U + F - W] + \lambda\Delta U^T\Delta U\} \qquad (61)$$

By setting the first derivative of J_d with respect to ΔU, equal to 0, we obtain

$$\partial J_d/\partial\Delta U = \lambda\Delta U + G^T[G\Delta U + F - W] \qquad (62)$$

Then the optimal control is given by

$$\Delta U = [G^TG + \lambda I]^{-1}G^T\{F - W\} \qquad (63)$$

To compute the control, we shall suppose that the future control variation is equal to zero.

$$\Delta u(t+i) = 0 \text{ for } i \geq 1 \qquad (64)$$

This assumption is made by analogy with a car driver, where the output is the car position with regard to the road, and the control variable is the angle made by the

steering-wheel with a fixed point on the dash board. It follows that

$$\Delta u(t) = G^T[F - W]/(g^Tg + \lambda) \qquad (65)$$

where g is the first row of the matrix G. Consequently, the control calculation does not need any matrix inversion which generally leads to numerical problems.

For the example mentioned above, $g^Tg = 2.0324$

<u>Remarks</u>

1. For a prediction horizon equal to the time delay we obtain the generalized minimum variance controller (Clarke and Gawthrop, 1979).

2. For oscillatory and unstable plants, the assumption (29) may be relaxed as follows:

$$\Delta u(t+i) = 0 \text{ for } i \geq Nu \geq 1$$

where Nu is the control horizon prediction.

In the following sections we shall introduce some improvements of the generalized predictive control algorithm by respectively considering generalized output and general disturbance process.

6.2. GENERALIZED OUTPUT

To obtain smoothness in the control signal variation,we can oblige the process output to follow the reference sequence output w(.) with user-specified dynamics. The latter may be defined as follows:

$$y_g(t) = M_r(q^{-1})y(t) \tag{66}$$

where $M_r(q^{-1})$ is a stable transfer function

Then the cost function will be expressed as follows:

$$J = \mathbf{E} \left[\left[\sum_{j=k}^{N_y} \{(y_g(t+j) - w(t+j))^2 + \lambda \; \Delta u(t+j-k)^2\} \right]/t \right] \qquad (67)$$

Notice that the difference $[y_g(t) - w(t)]$ represents the tracking error. For no steady state error, $M_r(1)$ must be equal to 1.

Consider the generalized output $y_g(.)$ at time $t + j$

$$y_g(t + j) = \{q^{-k}M_r(q^{-1})B(q^{-1})/A(q^{-1})\}u(t + j) +$$

$$+ \{M_r(q^{-1})/\Delta(q^{-1})A(q^{-1})\}e(t + j) \qquad (68)$$

As in paragraph 5.1, let us introduce the following identity

$$M_r(q^{-1}) = E_j(q^{-1})\Delta(q^{-1})A(q^{-1}) + q^{-j}F_j(q^{-1})/M_{rd}(q^{-1}) \qquad (69)$$

where $M_{rd}(q^{-1})$ is the denominator of the transfer function $M_r(q^{-1})$.

The optimal control strategy can be carried out in the same manner as in the preceding section. The only difference lies in the initialization of the identity (20). This Identity resolution must be started by the following initial polynomials $E_1(q^{-1})$ and $F_1(q^{-1})$.

$$E_1(q^{-1}) = M_r(0) = M_{rn}(0)/M_{rd}(0)$$

$$F_1(q^{-1}) = q[M_{rn}(q^{-1}) - E_1(q^{-1})A(q^{-1})\Delta(q^{-1})M_{rd}(q^{-1})] \qquad (70)$$

<u>Example 5:</u>

Consider the same system as in example 1 with

$M_r(q^{-1}) = 2.5 - 1.5q^{-1}$

Then for $j = 1$ to 2 we obtain the following polynomials $E_j(q^{-1})$ and $F_j(q^{-1})$.

$E_1(q^{-1}) = 2.5$ $\qquad\qquad$ $F_1(q^{-1}) = -1.25+1.9q^{-1}+0.35q^{-2}$

$E_2(q^{-1}) = 2.5-1.25q^{-1}$ $\qquad$ $F_2(q^{-1}) = 1.775-0.6q^{-1}-0.175q^{-2}$

6.3. GENERAL DISTURBANCE PROCESS

A general disturbance process may be obtained as follows:

$$e(t)/\Delta(q^{-1})=F_f(q^{-1})\varsigma(t)/\Delta(q^{-1}) \tag{71}$$

where $F_f(q^{-1})$ is an Hurwitz polynomial and $\varsigma(t)$ is uncorrelated stochastic sequence with zero mean and finite variance.

The plant model (equation 17) may be rewritten as follows:

$$A(q^{-1})\,\Delta(q^{-1})\,y(t+j) = B(q^{-1})\Delta(q^{-1})u(t+j-k) + F_f(q^{-1})\varsigma(t) \tag{72}$$

Let us introduce the following identity (as in the previous section)

$$F_f(q^{-1}) = E_j(q^{-1})A(q^{-1})\Delta(q^{-1}) + q^{-j}F_j(q^{-1}) \tag{73}$$

which may be solved in the same manner as that given by equation (20). The single difference lies in the initial conditions which are:

$$E_1(q^{-1}) = F_f(0) \quad ; \quad F_1(q^{-1}) = q[F_f(q^{-1}) - E_1(q^{-1})\Delta(q^{-1})] \tag{74}$$

Using identity (73) gives:

$$y(t+j) = q^{-k}EB_j(q^{-1})\Delta u^f(t+j) + F_j(q^{-1})y^f(t) + E_j(q^{-1})\,\xi(t+j) \qquad (75)$$

The optimal prediction $y^*(t + j\,/t)$ of the output $y(t + j)$ and the error prediction are given by:

$$y^*(t + j/t) = q^{-k}EB_j(q^{-1})\Delta u^f(t + j) + F_j(q^{-1})y^f(t) \qquad (76)$$

$$e_p(t + j) = E_j(q^{-1})\,\xi(t + j)$$

where $u^f(t + j)$ and $y^f(t + j)$ are the filtered input by $\{1/F_f(q^{-1})\}$, and $EB_j(q^{-1}) = E_j(q^{-1})B(q^{-1})$.

To determine the optimal control which minimizes the cost function (18), we have to express the output $y(t + j)$ or its prediction in terms of non-filtered input signal. To do this, it is necessary to express the term $\{EB_j(q^{-1})\}$ as a $F_f(q^{-1})$ function by dividing it by $F_f(q^{-1})$, say

$$EB_j(q^{-1}) = G^\circ_{j-k+1}(q^{-1})F_f(q^{-1}) + q^{-j+k-1}G^*_{j-k+1}(q^{-1}) \qquad (77)$$

where

$$G^\circ_j(q^{-1}) = g^\circ_{0,j} + g^\circ_{1,j}q^{-1} + \ldots + g^\circ_{nf+1-n-j,j}q^{nf+1-n-j}$$

$$G^*_j(q^{-1}) = g^*_{0,j} + g^*_{1,j}q^{-1} + \ldots g^*_{n-1,j}q^{-n+1} \qquad (78)$$

with degree de $F_f(q^{-1})$ equal to n_f

The output prediction is then given in terms of the filtered input-output and the non- filtered input as

$$y^*(t + j/t) = G^\circ_{j-k+1}(q^{-1})\Delta u(t + j - k) + G^*_{j-k+1}(q^{-1})\Delta u^f(t - 1) +$$

$$+ F_j(q^{-1})y^f(t) \qquad (79)$$

The terms $\{ G^*_{j-k+1}(q^{-1})\Delta u^f(t-1) + F_j(q^{-1})y^f(t)\}$ can be computed from the available data. For $j = k$ to Ny, equation (44) leads to the following matrix equation:

$$Y^* = G\Delta U + F \qquad (80)$$

where

$$Y^* = [y^*(t + k/t),\ldots\ldots\ldots, y^*(t + Ny/t)]^T$$

$$\Delta U = [\Delta u(t),\ldots\ldots\ldots, \Delta u(t + Ny - k)]^T$$

The vector F contains all the available data, and the matrix G is composed of the $G^\circ_{j-k+1}(q^{-1})$ polynomial coefficients.

Introducing (44) and (45) in (18) gives

$$J = \mathbf{E}\{[G\Delta U + F + E_p - W]^T[G\Delta U + F + E_p - W] + \lambda\Delta U^T\Delta U]/t\} \qquad (81)$$

where $E_p = [e_p(t + k),\ldots\ldots\ldots, e_p(t + Ny)]^T$

$$W = [w(t + k),\ldots\ldots\ldots\ldots, w(t + Ny)]^T$$

The prediction error is uncorrelated with the available data $\{y(t), y(t - 1),\ldots; \Delta u(t - 1), \Delta u(t - 2)\ldots; y^f(t),\ldots\ldots; u^f(t - 1),\ldots\ldots\}$ and by assumption the future variations of the control signal, then the control strategy will be derived from the optimization of the following deterministic function

$$J_d = [G\Delta U + F - W]^T[G\Delta U + F - W] + \lambda\Delta U^T\Delta U \qquad (82)$$

Setting the first derivative of J_d with respect to ΔU equal to zero, one obtains

$$\Delta U = [G^TG + \lambda I]^{-1}G^T[F - W] \qquad (83)$$

By supposing the future variation of the control to be equal to zero, we obtain an analogous expression to (30) for the optimal control.

Example 6:

Consider the same system as in example 1 and a filter with the following discrete transfer function: $F_f(q^{-1}) = 1 - 0.3q^{-1}$

For $j = 1$ to 3 the polynomial $E_j(q^{-1})$, $F_j(q^{-1})$, $G^*_j(q^{-1})$ and $G^\circ_j(q^{-1})$ are listed below

$$E_1(q^{-1}) = 1 \; ;$$

$$F_1(q^{-1}) = -0.2 + 0.76q^{-1} + 0.14q^{-2}$$

$$E_2(q^{-1}) = 1 - 0.2q^{-1}$$

$$F_2(q^{-1}) = 0.74 - 0.012q^{-1} - 0.028q^{-2}$$

$$E_3(q^{-1}) = 1 - 0.2q^{-1} + 0.74q^{-2}$$

$$F_3(q^{-1}) = 0.062 + 0.5344q^{-1} + 0.1036q^{-2}$$

$$G^*_1(q^{-1}) = 0.8$$

$$G^\circ_1(q^{-1}) = 1$$

$$G^*_2(q^{-1}) = 0.08$$

$$G^\circ_2(q^{-1}) = 1 + 0.6q^{-1}$$

$$G^*_3(q^{-1}) = 0.616$$

$$G^\circ_3(q^{-1}) = 1 + 0.6q^{-1} + 0.82q^{-2}$$

An improved scheme bringing together the advantages of the predictive control and pole placement techniques has been proposed by Irving et al. (1986) . The pole placement concept is motivated by simplicity of performance definition. The resulting algorithm will be presented in the next section.

7. LONG-RANGE PREDICTIVE CONTROL

Among the adaptive control algorithms developed in the last decade (Åström, 1983; Seborg et al. 1986) the long-range predictive control approach seems to be very attractive and less sensitive to plant model mismatch. An enhancement version of the generalized predictive control algorithm is described in the following.

Let us consider a disturbed plant which can be modeled by the following linear discrete system:

$$A(q^{-1})y(t) = q^{-k} B(q^{-1})u(t) + v(t) \qquad (84)$$

where $y(t)$, $u(t)$ and $v(t)$ denote the ouput, the intput and the disturbance. k is the time delay expressed as an integer multiple of the sampling period ($k \geq 1$).

The disturbance will be represented in discrete time as follows:

$$D(q^{-1}) v(t) = [F(q^{-1})/G(q^{-1})] e(t) \qquad (85)$$

where $\{e(t)\}$ is a stochastic noise variable (random variable with normal distribution and zero mean). $A(q^{-1})$, $B(q^{-1})$, $D(q^{-1})$,

$F(q^{-1})$ and $G(q^{-1})$ are polynomials in the backward shift operator q^{-1} of order n_a, n_b, n_d, n_f and n_g respectively.

The $A(q^{-1})$ and $B(q^{-1})$ polynomials represent the nominal plant response; while the $F(q^{-1})$, $G(q^{-1})$ and $D(q^{-1})$ polynomials characterize the unmodelled plant response. $F(q^{-1})$ and $G(q^{-1})$ are asymptotically stable polynomials and $D(q^{-1})$ denotes the internal model of external disturbances. More particularly, the design polynomial $D(q^{-1})$ allows us to introduce an offset-free disturbance compensation; e.g., the choice $D(q^{-1}) = 1 - q^{-1}$ provides a natural integral action in the control law. On the other hand the $F(q^{-1})$ and $G(q^{-1})$ polynomials should be chosen such that the filter $\{F(q^{-1})/G(q^{-1})\}$ is a low-pass one. By the way, the same filter may be used in the parameter estimation procedure to reduce the high frequency modes due to both noise and unmodelled dynamics. This ensures smooth variations of parameter estimates. The latter property is quite coherent with the stability requirements of adaptive controllers.

The controller will be designed to achieve the following objective:

$$P(q^{-1})[y(t) - \{B(q^{-1})/ B^*(q^{-1})\} w(t)] = b(t) \qquad (86)$$

where $P(q^{-1})$ is an asymptotically stable polynomial specifying the desired regulation dynamics. The sequence $w(.)$ characterizes the desired tracking dynamics. $b(t)$ is a bounded zero mean sequence. $B^*(q^{-1})$ will be chosen so as to obtain a static gain equal to 1 (no steady-state error). This polynomial will be

considered in the following as a constant
($B^*(q^{-1}) = b^*$; $B^*(q^{-1}) = B(1)$ if $B(1) \neq 0$ and $B^*(q^{-1}) = 1$ otherwise).

The control objective (86) can be achieved by the pole placement controller which has the general form shown in figure 9.

$$D(q^{-1})\,S(q^{-1})u(t) + R(q^{-1})y(t) = \{P(q^{-1})/b^*\}w(t+k) \tag{87}$$

where $R(q^{-1})$ and $S(q^{-1})$ are asymptotically stable polynomials. They are determined by requiring that

$$P(q^{-1}) = A(q^{-1})D(q^{-1})S(q^{-1}) + q^{-k}B(q^{-1})\,R(q^{-1}) \tag{88}$$

which is known as the Bezout identity (Diophantine equation).

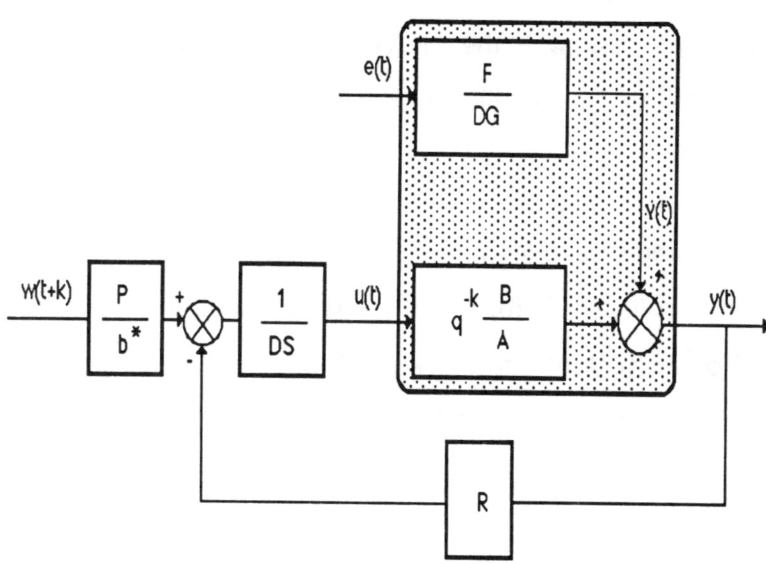

Figure 9. Block diagram of the control law of (87)

The solution of equation (88) is unique (Kucera, 1979; Åström and Wittenmark, 1984) if and only if:

- the least greatest common divisor of the polynomials $A(q^{-1})$ and $B(q^{-1})$ divide the polynomial $P(q^{-1})$

- deg R = deg A+ deg D - 1

- deg S = deg B + k -1 $\qquad\qquad$ (89)

- deg P $\leq$ deg R+ deg S + 1

where deg denotes the polynomial degree.

Conditions (88) and (89) constitute the disadvantage of pole placement techniques, especially in an adaptive context (Clarke, 1982). In particular the pole placement design is more sensitive to model order choice.

We shall deal with the above-mentioned control objective (equation 86) using the long range predictive design as in Clarke et al. (1984).

Simple algebraic manipulations of equations (84 - 86) result in the following closed-loop input-output relations:

$$P(q^{-1})[y(t) - \{B(q^{-1})/b^*\}w(t)] = \{S(q^{-1})F(q^{-1})/G(q^{-1})\}\, e(t) \qquad (90)$$

$$D(q^{-1})P(q^{-1})[u(t)-\{A(q^{-1})/b^*\}w(t+k)]=-\{R(q^{-1})F(q^{-1})/G(q^{-1})\}e(t) \qquad (91)$$

To do this let us introduce the following conceptual variables:

$$\epsilon_y(t) = P(q^{-1})[y(t) - \{B(q^{-1})/b^*\}w(t)] \qquad (92)$$

$$\epsilon_u(t) = P(q^{-1})[u(t) - \{A(q^{-1})/b^*\}w(t+k)] \qquad (93)$$

which may be referred to respectively as the tracking input error and the tracking output error.

The control objective (equation 86) may be restated in terms of cost function minimization as follows:

$$J = \mathbf{E}\left[\left\{\sum_{i=k}^{Ny} \left[\{\epsilon_y(t+i)\}^2 + \lambda \{D(q^{-1})\,\epsilon_u(t+i-k)\}^2\right]\right\}/t\right] \qquad (94)$$

under the basic assumption that:

$$D(q^{-1})\,\epsilon_u(t+j) = 0 \text{ for } j \geq Nu \qquad (95)$$

where Nu is the control horizon.

The assumption (95) constitutes an advantage of the control algorithm presented here over the G.P.C. In fact, the assumption made by Clarke et al. (1984) that $(1 - q^{-1})u(t + j) = 0$ for $j > Nu$ is valid only if the desired output is constant.

The parameter λ can be used to trade off the variance of the error $\epsilon_y(t)$ against the variance of the error $\epsilon_u(t)$.

Based on equations (55) and (56), the tracking errors may be expressed as follows:

$$\epsilon_y(t) = \{S(q^{-1})F(q^{-1})/G(q^{-1})\}\, e(t) \qquad (96)$$

$$D(q^{-1})\,\epsilon_u(t) = -\{R(q^{-1})F(q^{-1})/G(q^{-1})\}\, e(t) \qquad (97)$$

Equations (84), (85), (92) and (93) lead to the following model reparametrization (M'Saad et al., 1986):

$$A(q^{-1})\{D(q^{-1})\epsilon_y(t)\} = q^{-k}B(q^{-1})\{D(q^{-1})\epsilon_u(t) +$$

$$+ \{P(q^{-1})F(q^{-1})/G(q^{-1})\}e(t) \qquad (98)$$

Let us derive the optimal predictor (in least squares sense) of $\epsilon_y(t+j)$, given information up to time t, for the model (98). It is always possible to find $S_j(q^{-1})$ and $R_j(q^{-1})$ polynomials which satisfy the polynomial identity:

$$[P(q^{-1})F(q^{-1})/G(q^{-1})] = A(q^{-1})D(q^{-1})S_j(q^{-1}) + q^{-j}R_j(q^{-1})/G(q^{-1}) \quad (99)$$

where deg $S_j = j - 1$ and deg $R_j = \max\{\text{deg } A + \text{deg } D + \text{deg } G - 1$, deg F + deg P - j\}$

Combining equations (98) and (99) leads to:

$$\epsilon_y^*(t + j/t) = \{BGS_j /PF\} D\epsilon_u(t + j - k)\} + \{R_j/PF\} \epsilon_y(t) \quad (100)$$

$\epsilon_y^*(t + j/t)$ indicates the conditional expectation value of the error $\epsilon_y(.)$ at time $t + j$, based on the data available at time t.

As the cost function (equation 94) should be minimized with respect to the sequence $\{D(q^{-1}) \epsilon_u(t+i) \}$ for $i = 0$, Nu - 1, these components have to appear explicitly in the prediction equation (100). To do so we will consider the polynomial identity:

$$\{B(q^{-1})G(q^{-1})\}S_j(q^{-1}) = P(q^{-1})F(q^{-1})L_{j-k+1}(q^{-1}) +$$

$$+ q^{k-j-1}K_{j-k+1}(q^{-1}) \quad (101)$$

with:

$$\text{deg } L_{j-k+1} = j - k$$

$$\text{deg } K_{j-k+1} = \max(\text{deg } P + \text{deg } f - 1, \text{deg } B + \text{deg } G + k - 2) \quad (102)$$

Combining equations (100) and (101) gives:

$$\epsilon_y^*(t + j/t) = L_{j-k+1} D \, \epsilon_u(t + j - k) + \rho(t,j) \tag{103}$$

The control sequence has been broken into past and future elements. The term $\rho(t,j)$ is equal to:

$$\rho(t,j) = \{R_j/PF\}\epsilon_y(t) + \{K_{j-k+1}/PF\}D\,\epsilon_u(t-1) \tag{104}$$

The control criterion may be written as:

$$J = [\, G \, DE_u + R]^T[\, G \, DE_u + R] + \lambda[DE_u]^T[DE_u] \tag{105}$$

where:

$$E_u = [\epsilon_u(t)\,,\,\epsilon_u(t+1)\,,\,\ldots\,,\,\epsilon_u(t + Ny - 1)]^T$$

$$R = [\rho(t\,,\,k)\,,\,\rho(t\,,\,k+1)\,,\,\ldots\,,\,\rho(t\,,\,Ny)]^T$$

$$G = \begin{array}{ccccccc} l_0 & 0 & 0 & . & . & . & 0 \\ l_1 & l_0 & 0 & . & . & . & 0 \\ . & . & . & . & . & . & . \\ l_{Ny-k} & & . & . & . & . & l_{Ny-Nu-1} \end{array} \tag{106}$$

The parameters l_i are the L_{j-k+1} polynomial coefficients.

J is minimized relative to DE_u by cancelling its gradient. The gradient of J is cancelled by the following choice:

$$DE_u^\circ = -[\, G^T G + \lambda I]^{-1}[\, G^T R \,] \tag{107}$$

As the control is to be implemented in the receding horizon sense, one has:

$$D\,\epsilon_u(t) = [-1\ 0\ \ldots\ 0]\,[\,G^T G + \lambda I]^{-1}[\,G^T R\,] \tag{108}$$

The control law is hence given by:

$$D(q^{-1})u(t) = \{D(q^{-1})/P(q^{-1})\}\,\epsilon_u(t) + \{D(q^{-1})A(q^{-1})/b^*\}w(t+k) \tag{109}$$

<u>Example 7:</u>

Consider a system given by (49-50), where:

$$A(q^{-1}) = 1 + 0.3q^{-1} - 0.25q^{-2}\ \ ;\ \ B(q^{-1}) = 1 + 0.5q^{-1}\ \ ;\ \ k = 2$$
$$P(q^{-1}) = 1 + 1.2q^{-1} + 0.47q^{-2} + 0.06q^{-3}\ \ ;\ \ D(q^{-1}) = 1 - q^{-1}$$
$$F(q^{-1}) = 1 + 0.3q^{-1}\ \ ;\ \ G(q^{-1}) = 1.$$

For $j = 1$ to 4 ($Ny = 4$) we obtain the following expression for the polynomials: R_j, S_j, L_j and K_j.

$$S_1(q^{-1}) = 1\ ;\ S_2(q^{-1}) = S_1(q^{-1}) + 2.2\ q^{-1}\ ;$$

$$S_3(q^{-1}) = S_2(q^{-1}) + 2.92\ q^{-2}\ ;\ S_4(q^{-1}) = S_3(q^{-1}) + 3.205\ q^{-3}$$

$$R_1(q^{-1}) = 2.2\ + 1.38\ q^{-1} - 0.0489\ q^{-2} + 0.018\ q^{-3}$$

$$R_2(q^{-1}) = 2.92\ + 1.161\ q^{-1} - 0.532\ q^{-2} + 0.018\ q^{-3}$$

$$R_3(q^{-1}) = 3.205\ + 1.074\ q^{-1} - 0.074\ q^{-2} + 0.018\ q^{-3}$$

$$R_4(q^{-1}) = 3.3175 + 1.05\ q^{-1} - 0.783\ q^{-2} + 0.018\ q^{-3}$$

$$K_1(q^{-1}) = 1.5 + 0.63\ q^{-1} - 0.0599\ q^{-2}$$

$$K_2(q^{-1}) = 1.75\ + 0.69\ q^{-1} - 0.089\ q^{-2}$$

$$K_3(q^{-1}) = 1.8\ + 0.69\ q^{-1} - 0.105\ q^{-2}$$

$$L_1(q^{-1}) = 1. \; ; \; L_2(q^{-1}) = L_1(q^{-1}) + 1.5\, q^{-1} \; ;$$

$$L_3(q^{-1}) = L_2(q^{-1}) + 1.75\, q^{-2}$$

The following program computes the solution of the first identity (equation 63) involved in the modified generalized predictive control algorithm.

```
      SUBROUTINE RESDI1(AB,NAB,PF,NPF,R,NR,S,NS,RI,J)
c     AB(q-1)= A(q-1)D(q-1)G(q-1)
c     NAB   degree of the polynomial AB(q-1); i.e. na+nd+ng
c     PF(q-1)= P(q-1)F(q-1)
c     NPF   degree of the polynomial PF(q-1); i.e. np+nf
c     The vectors AB,PF,R,S contain the coefficients of the
c     polynomials AB(q-1), PF(q-1),
c      Rj(q-1) and Sj(q-1) respectively
c     RI is a working vector
c
      DIMENSION AB(20),PF(20),R(20),S(20),RI(20)
c
c     NR=MAX{deg AB-1,deg PF-J}
c
      NR=NPF-J
      NS=J-1
      IF((NAB-1).GT.(NPF-J)) NR=NAB-1
```

```
      IF(J.ne.1) GO TO 4
      S(1)=PF(1)
      DO 5 i=1,NR+1
      i1=i+1
      R(i)=PF(i1)-S(1)*AB(i1)
  5   CONTINUE
      GO TO 10
  4   RR=R(1)
      DO 7 i=1,NR+1
      i1=i+1
  7   RI(i)=R(i1)-RR*AB(i1)
      DO 8 i=1,NR+1
  8   R(i)=RI(i)
      S(J)=RR
 10   CONTINUE
      RETURN
      END
```

The following program performs the numerical resolution of the second identity (equation 101) involved in the design of the adaptive control algorithm presented.

```
      SUBROUTINE RESDI2(B,P,S,PL,PK,NPK,JR)
c
c     The vectors  B, P, S, PL, PK contain the coefficients of the
c     polynomials B(q^-1), P (q^-1),
```

```
c      Sj(q⁻¹), Lj-k+1(q⁻¹) and Kj-k+1(q⁻¹) respectively
c      NPK   degree of the polynomial  Ki(q⁻¹)
c      JR    = j - k +1  ( k represents the time delay)
c
       DIMENSION B(20),P(20),S(20),PL(20),PK(20)
c
       IF(JR.EQ.1)  PL(jr)=B(1)*S(1)/P(1)
       IF(JR.NE.1)  THEN
       PL(JR)=0.0
       DO 1 i=1,JR
       i1=JR+1-i
  1    PL(JR)=PL(JR)+B(i)*S(i1)
       DO 2 i=1,JR-1
       i1=JR+1-i
  2    PL(JR)=PL(JR)-PL(i)*P(i1)
       PL(JR)=PL(JR)/P(1)
       END IF
       DO 3 i=1,NPK+1
       PK(i)=0.0
       DO 4 i1=1,i+JR+1
       i2=i+JR+1-i1
  4    PK(i)=PK(i)+B(i1)*S(i2)
       DO 5 i3=1,JR
  5    PK(i)=PK(i)-PL(i3)*P(JR+i+1-i3)
  3    CONTINUE
```

```
RETURN
END
```

In this part of this chapter we have presented the most robust adaptive techniques, namely the generalized predictive algorithm and its improvement developed by Irving et al. (1986) and M'Saad et al. (1986). This control approach allows one to specify in a straightforward manner the desired tracking dynamics. However the design parameters (P,G,F,D,Ny,Nu) have to be chosen in order to get an asymptotically stable regulation dynamics. Indeed, the latter is no longer given by the polynomial $P(q^{-1})$ because of the cost function minimization. On the other hand the involved control algorithm is as easy to implement as the standard G.P.C.

The second part of this chapter is dedicated to the learning control approach.

8. INTRODUCTION TO LEARNING CONTROL

A learning system works in a random environment. Based on the environment responses, it improves its behaviour (In the control, the environment models the controlled process).

To do this, the learning automaton adjusts its probability distribution in such a

way as to minimize some criteria. It consists of a performance evaluation unit and a stochastic automaton with variable structure (figure 10).

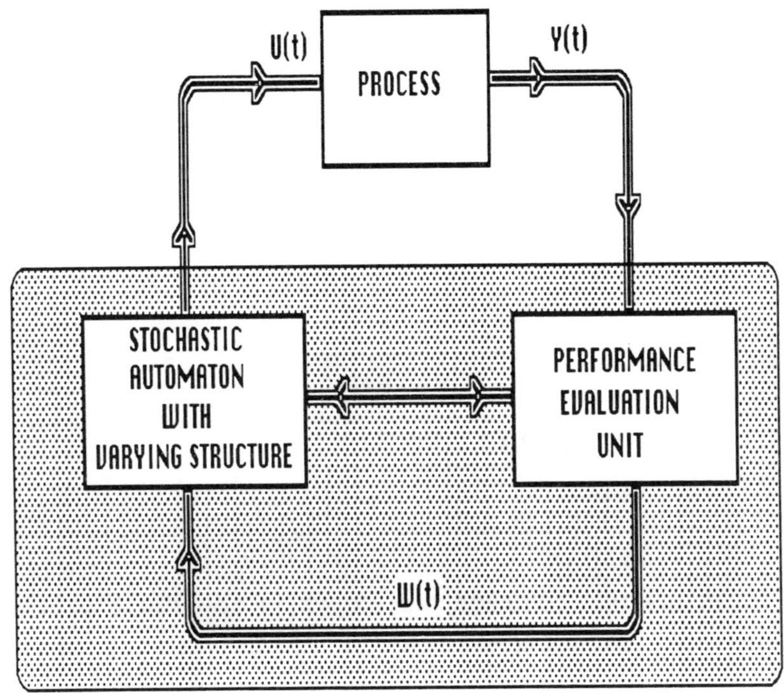

Figure 10. Learning automaton

A stochastic automaton is defined as a system: $\{W,\varnothing,\Pi,P,R,G\}$. where:

- W(t) is the response of the performance system (W(t) = 1 is called "penalty" and

W(t) = 0 is called the "non penalty" or "reward")

- $\varnothing = \{\varnothing_1,. .. ,\varnothing_S\}^T$ is the set of internal states.

- $\Pi = \{\pi_1,. .. , \pi_r\}^T$ with $r \leq s$ is the output or the action set. The action chosen by the automaton plays the role of the environment input (the control variable).

- $P(t) = \{p_1(t),. .. , p_N(t)\}^T$ is the state probability distribution at time t.

If the automaton chooses the action $\pi(t)$ at time t (t = 1, 2,. ..) then:

$$p_i(t) = Pr \{\pi(t) = \pi_i \}$$

and

$$\sum_{i=1}^{N} p_i(t) = 1 \quad \text{for all t}$$

(110)

To ensure this condition the projection operator (Poznyak, 1975) is sometimes used.

- R: the learning algorithm (reinforcement scheme or updating scheme) which changes the probability vector from P(t) to P(t+1) { P(t+1) = R[P(t), W(t)] }. It can be written as follows:

$$p_i(t+1) = p_i(t) + g_i[t , p_1(t), p_2(t),......, p_N(t)]$$

(111)

$$i = 1,...., N \; ; t = 1, 2,.......$$

where $g_i(.)$ is an updating term for $p_i(.)$ at the stage of operation of the automaton-environment combination. To preserve probability measure, it is necessary that the following condition be satisfied for all t:

$$\sum_{i=1}^{N} g_i[t, p_1(t), p_2(t),......, p_N(t)] = 0 \qquad (112)$$

If the function $g_i(.)$ is linear (or non-linear), the reinforcement scheme is said to be linear (or non-linear).

- G is the mapping of the set $\emptyset$ onto the set Π (G: $\emptyset \mapsto \Pi$). In general, G may be a stochastic function. It is often assumed that G is deterministic and one-to-one (i.e. $r = s = N$) and $s < \infty$ (Narendra and Thathachar, 1974 ; Lakshmivarahan, 1981).

The environment (random medium) emits output $w(t) = 1$ with probability c_i ($i = 1,.. , N$) and output $w(t) = 0$ with probability $1 - c_i$. The parameters c_i ($i = 1,.. , N$) represent the probability of penalization. It is assumed that the c_i are unknown initially. The problem will be trivial if they are known a priori.

The control variable $u(t)$ varies in the domain $[u_{min}$, $u_{max}]$. This domain will be discretized into a set of N^N values i.e. (u_i , $i = 1 , N^N$). The number of internal states will be considered equal to the number of the set action ($s = N$).

At every time (sampling period) the automaton chooses an action on the basis of a probability distribution defined for only N values ($i = 1,N$). Then, the system performance reaction (penalty, non penalty or reward) and a reinforcement scheme are used to update the probability distribution.

The system performance generates either reward or penalty according to the control objective.

8.1. LEARNING AUTOMATA: DEFINITIONS AND PROPERTIES

We shall present some definitions and properties of learning stochastic automata which characterize the environment and the behaviour of the learning automaton. The environment corresponds to the medium where the automaton operates. The automaton has variable structure.

Environment

Stationarity: When the probabilities c_i ($i = 1,...N$) are constant, the environment is stationary. Otherwise, the medium is said to be non-stationary.

Automaton

Let $A_p(t)$, defined by:

$$A_p(t) = \mathbf{E}[w(t)/P(t)] = \sum_{i=1}^{N} p_i(t)c_i \qquad (113)$$

be the average penalty received by the automaton. If the

actions u_i are selected with equal probability $\{p_i(t) = 1/N$; $i = 1,2,...,N\}$, the obtained average penalty is denoted by A_p^* and given by:

$$A_p^* = \{\sum_{i=1}^{N} c_i\}/N \tag{114}$$

*** expediency**

The learning automaton is expedient if

$$\lim_{t \mapsto \infty} E[A_p(t)] < A_p^* \tag{115}$$

***optimality**

When the learning automaton is such that:

$$\lim_{t \mapsto \infty} E[A_p(t)] < c_{min} \tag{116}$$

with $c_{min} = \min_i [c_i]$ $i = 1,2,.......,N$

it is called optimal. Optimality means that the minimum penalty probability is selected with probability one when $t \mapsto \infty$.

*** ϵ-optimality**

A learning automaton has an ϵ-optimality property if the following condition:

$$\lim_{t \mapsto \infty} E[A_p(t)] < c_{min} + \epsilon \tag{117}$$

is verified for any arbitrary $\epsilon > 0$. This parameter indicates how close the performance of the learning automaton is to optimality.

8.2. HIERARCHICAL STRUCTURE OF AUTOMATON

To obtain a high quality of level control, the set of the action must be increased. Consequently, the behaviour of the automaton will be slow, and the necessary memory needed by the implementation of the learning control algorithms will also increase. To avoid these problems, we can use a hierarchical system of automata which is at different

levels composed of a single automaton with limited number of actions (N). This approach was developed by Thathachar and Ramakrishnan (1981).

The first level consists of a single automaton, the second one is composed of N single automata and the ith level is formed by (N^{i-1}) automata (figure 11).

Let us consider a hiearchical system of automata with N levels.

The control variable domain variation will be discretised in (N^N) intervals.

If the actions chosen at the various levels are:

$$\pi_{j1} \, , \pi_{j1j2} \, , \pi_{j1j2j3} \, , \cdots , \pi_{j1j2....jN}$$

$$(j1, j2 ,......, jN \; ; \quad ji = 1,2,...,N)$$

The corresponding automata are: $A, A_{j1}, A_{j1j2},..., A_{j1j2...jN-1}$

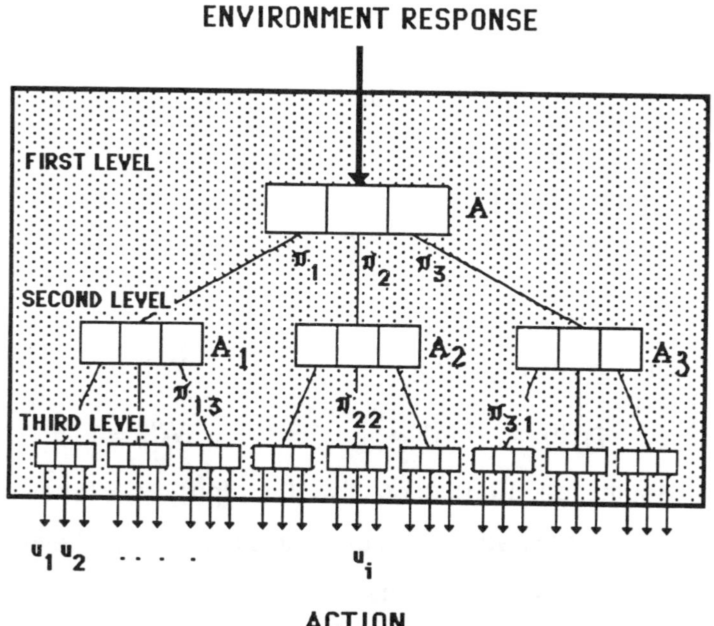

Enlargement :

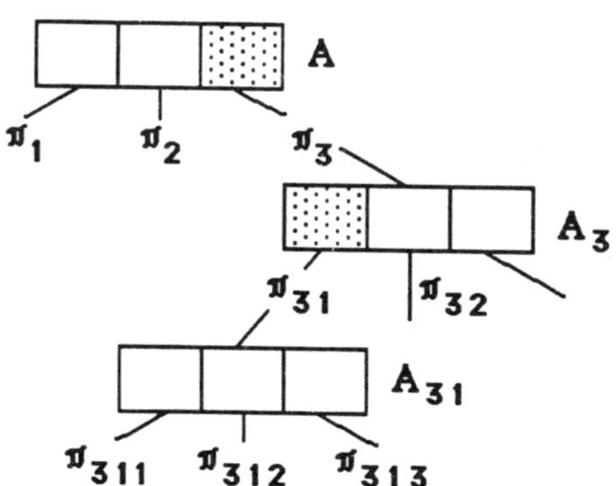

Figure 11. Hierarchical system of learning automata

If the action selected at instant t by the first level automaton is π_{j1}, the automaton concerned is A, and the probability distribution will be adapted by the following rule (Thathachar and Ramakrishnan, 1981):

$$p_{j1}(t + 1) = p_{j1}(t) + \mu\{(p(t)\}[1 - p_{j1}(t)] \qquad , \quad W(t) = 0$$

$$= p_{j1}(t) \qquad , \quad W(t) = 1$$

$$p_{i1}(t + 1) = p_{i1}(t)[1 - \mu\{p(t)\}] \qquad , \quad W(t) = 0 \qquad (118)$$

$$= p_{i1}(t) \qquad i1 \neq j1 \qquad , \quad W(t) = 1$$

By assumption, $\mu\{p(t)\}$ is positive $(0 < \mu\{p(t)\} < 1)$, for all $p(t)$.

To carry out the adaptation mechanism for the nth level, let us introduce the following notations.

Let us denote $i1i2....in$ by In, and $j1j2....jn$ by Jn.

At the nth level, if the selected action at time t is π_{Jn}, the concerned automaton will be noted A_{Jn-1}.

The probability distribution is adjusted by the following algorithm:

$$p_{Jn}(t + 1) = p_{Jn}(t) + \mu_{Jn-1}\{ p(t)\}[1 - p_{Jn}(t)] \qquad , \quad W(t) = 0$$

$$= p_{Jn}(t) \qquad , \quad W(t) = 1$$

$$(119)$$

$$p_{Jn-1in}(t + 1) = p_{Jn-1in}(t)[1 - \mu_{Jn-1}\{p(t)\}] \qquad , \quad W(t) = 0$$

$$= p_{Jn-1in}(t) \qquad in \neq Jn \qquad , \quad W(t) = 1$$

For all other actions

$$p_{In}(t+1) = p_{In}(t) \qquad , \quad W(t) = 0 \text{ or } 1. \qquad (120)$$

A $\mu\{.\}$ function is associated with each automaton in the hierarchy.

The following algorithm (Thathachar and Ramakrishnan, 1981), which ensures ϵ-optimality, is chosen to adapt the $\mu\{.\}$ function.

$$\mu_{Jn}\{t\} = \mu_{Jn-1}\{t\} / p_{Jn}\{t+1\} \qquad (121)$$

Based on the probability distribution, the first level automaton generates an action (π_{in}) randomly. This, in turn, activates the automaton A_{in} in the second level, which chooses an action ($\pi_{i1\ i2}$) from its action probability distribution. Consequently, the automaton $A_{i1\ i2}$ is activated. The final action selected at the last level generates a control action for the process to be controlled, and in turn, the response of the environment (the plant under consideration) is used to update the distribution probability at each level. This procedure will be repeated at every sampling period.

The vector probability P(t) at each level depends only on its previous value and on the action selected at the previous level.

Simple calculations can easily show that the reinforcement scheme presented above, which is a mapping that determines the vector probability distribution at time t + 1 based on the knowledge of this distribution at time t, is such that the sum of the vector probability distribution is equal to 1.

For every sampling period the control algorithm, based on a single or a hierarchical structure of automata, performs the following steps:

Data: Cchoose the initial setting of the probability distribution, the parameters of the reinforcement scheme and, of course, the number N of actions.

Step 1: Choice of one action π_i (related to the control action).

The technique used by the algorithm to select one action π_i among N possibilities is based on the generation of a normally distributed random variable Z (any specific machine subroutine: RANDU, etc. can be used to carry out the random variable Z). The algorithm chooses the action π_i such that i is equal to the least value of j, verifying the following constraint

$$\sum_{j=1}^{i} p_j(t) \geq Z \qquad (122)$$

Step 2 Application of this control to the controlled process (the control variable of the plant under consideration corresponds to the automaton output)

Step 3: Measurement of the controlled variable (s)

Step 4: Use of this measure (s) in the performance evaluation unit to decide, on the basis of heuristic and/or analytical rules, whether the choice of the action π_i complies with the conditions of a good process behaviour. This leads the unit to deliver a response which can correspond, in the first case, to a penalty or inaction (non-penalty) or, in the second case, to a reward.

Step 5: Use of this response to adjust the probability distribution associated to the set   of N actions by means of an optimal or ϵ-optimal reinforcement scheme which generates a new probability distribution $P(t+1)$ from the previous one $P(t)$ according to the value of $W(t)$.

- Return to step 1.

The implementation of a hierarchical structure of automata using a digital computer may be facilitated by the program HILEAR, which uses external random number generator.

```
      SUBROUTINE HILEAR(P,VP,N,39,MU,L1,L2,L3,W)
      DIMENSION P(N,39),VP(N)
C
C     W     PERFORMANCE EVALUATION UNIT RESPONSE
C     N     THE NUMBER OF ACTIONS ASSOCIATED TO ALL
C           SINGLE AUTOMATON
C     P     P(i,j*N) THE PROBABILITY VECTOR ASSOCIATED TO
```

```
c           THE jth AUTOMATON IN THE ith LEVEL (FOR N=3, WE
c           HAVE TO CONSIDER 39 COMPONENTS)
c     MU    THE COEFFICIENT μi
c     Li    i=1,N THE SELECTED ACTION IN THE iTH LEVEL BY
c           THE ACTIVATED  AUTOMATON
c
      CALL TREE(P,VP,N,39,L1,L2,L3)
      IF (W.EQ.1)  GOTO 1
      CALL ADAP(P,N,MU)
   1  RETURN
      END
c
c
      SUBROUTINE TREE(P,VP,N,39,L1,L2,L3)
      DIMENSION VP(N),P(N,39)
       Z= (A random variable 0 < Z < 1, generated by any program
          like RAND(.), RANDOM, etc.)
      L=1
      DO  10 i=1,N
  10  VP(i)=P(1,i)
      CALL LOCAII(VP,II,Z,N)
      L1=II
      Z= (..)
      DO  11 i=1,N
      i1=N*(L1-1)+i
```

```
   11  VP(i)=P(2,i1)
       CALL LOCALI(VP,II,Z,N)
       L2=II
       Z= (..)
       DO  12 i=1,N
       i2=N*N*(L1-1)+N*(L2-1)
   12  VP(i)=P(3,i2)
       CALL LOCALI(VP,II,Z,N)
       L3=II
       RETURN
       END
C
       SUBROUTINE LOCALI(VP,II,Z,N)
C
C   ACTION SELECTION
C
       DIMENSION VP(N)
       A=0.0
       DO  10 i=1,N
       A=A+VP(i)
   10  IF(A.GE.Z) GOTO 11
   11  II=i
       RETURN
       END
C
C
```

```
      SUBROUTINE ADAP(P,N,MU)
C
C     UPDATING OF ACTION PROBABILITIES
C
      DIMENSION P(N,39)
      L=1
      PP=P(L,L1)
      DO  10 i=1,N
   10 P(L,i)=P(L,i)*(1.0-MU)
      P(L,L1)=PP+MU*(1.0-PP)
      MU1=MU/P(L,L1)
      L=2
      J=N*(L1-1)+L2
      PP=P(L,J)
      NL21=N*(L1-1)+1
      NL2N=N*(L1-1)+N
      DO  11 i=NL1,NL2
   11 P(L,i)=P(L,i)*(1.0-MU1)
      P(L,J)=PP+MU1*(1.0-PP)
      MU2=MU1/P(L,J)
      L=3
      J=N*N*(L1-1)+N*(L2-1)+L3
      PP=P(L,J)
      NL31=N*N*(L1-1)+1+N*(L2-1)
      NL3N=N*N*(L1-1)+N*L2
      DO  12 i=NL31,NL3N
```

```
12  P(L,i)=P(L,i)*(1.0-MU2)
    P(L,J)=PP+MU2*(1.0-PP)
    RETURN
    END
```

To improve performance of the learning control system, it sems to be useful to adapt in real-time the number of actions of the automaton.

8.3. REAL TIME SELECTION OF THE NUMBER OF AUTOMATON ACTIONS

For control or optimization purposes, it is sometimes necessary to change the number of actions. For example, the optimal synthesis of industrial processes can be stated as a large mixed-integer programming problem, in which the discrete variables refer to the process structure (connection between unit operations, type of unit operation, etc.) and the continuous represent the operating conditions. This optimization problem can be solved using a learning automaton where a probability distribution is associated to the set of feasible solutions. This set of feasible solutions decreases with time. It is also necessary to change in real time the number of automaton actions.

We shall present a learning algorithm with a changing number of actions, which was developed by Thathachar and Harita (1987).

Let us denote by $S_i(t)$ i =1, Ns the ith subset of actions selected at time t, where if N represents the automaton action number, Ns (the total number of subsets of actions) is equal to:

$$Ns = N^2 - 1 \tag{123}$$

For example :

N = 2, leads to the following subsets of actions

$$S_1 = [\pi_1], S_2 = [\pi_2], S_3 = [\pi_1, \pi_2]$$

N = 3, leads to the following subsets of actions

$$S_1 = [\pi_1], S_2 = [\pi_2], S_3 = [\pi_3], S_4 = [\pi_1, \pi_2], S_5 = [\pi_1, \pi_3],$$

$$S_6 = [\pi_2, \pi_3], S_7 = [\pi_1, \pi_2, \pi_3]$$

The probability of choosing the ith subset of actions at time t will be noted

$$q_i(t) = Pr [S(t) = S_i(t)] \tag{124}$$

Of course

$$\sum_{i=1}^{Ns} q_i(t) = 1 \text{ for all } t \tag{125}$$

The associated probability vector will be noted Q(t), $(Q(t) = [q_1(t), q_2(t),..., q_{Ns}(t)]^T)$.

The probability $p^*_i(t)$ of chosing the ith action π_i from the selected subset $S_i(t)$ of actions ($p^*_i(t) = Pr [\pi(t) = \pi_i / S(t), \pi_i \in S(t)]$, $P^*(t) = [p^*_1(t), p^*_2(t),..., p^*_{Ns}(t)]^T$) is given by:

$$p^*_i(t) = p_i(t)/K(t) \tag{126}$$

where $K(t) = \sum p^*_i(t)$, $\pi_j \in S(t)$

The learning automaton with a changing number of actions may be described by the following steps:

At time t ,the control algorithm based on the learning automaton with a changing number of actions performs the following steps:

Data: choose the initial setting of the distribution of probabilities ($P(0)$ and $P^*(0)$), the parameters of the reinforcement scheme and of course the total number N of actions.

Step 1: choice of one subset $S_i(t)$. The technique used by the algorithm to select one subset $S_i(t)$ among Ns possibilities is based on the generation of normally distributed random variables Z (any specific machine subroutine: RANDU, etc., can be used to do this). The algorithm chooses the subset $S_i(t)$ such that i is equal to the least value of j, verifying the following constraint:

$$\sum_{j=1}^{i} p^*_j(t) \geq Z$$

Step 2: calculate the vector $P^*(t)$ components

$$p^*_j(t) = p_i(t)/K(t), \ \pi_j \in S(t)$$

Step 3: use the following scheme to adapt the probability

vector $P^*(t)$

$\underline{W(t) = 0}$

$$p^*_j(t + 1) = p^*_j(t) + \beta(1 - p^*_j(t)) \text{ for } \pi(t) = \pi_j$$

$$p^*_i(t + 1) = p^*_i(t) - \beta \, p^*_i(t) \text{ for } i \neq j \, , \, \pi(t) = \pi_i \tag{127}$$

where β is a constant positive number $\beta \in (0 , 1)$

$\underline{W(t) = 1}$

$$p^*_j(t + 1) = p^*_j(t) \text{ for } \pi(t) = \pi_j$$

$$p^*_i(t + 1) = p^*_i(t) \text{ for } i \neq j \, , \, \pi(t) = \pi_i \tag{128}$$

The other components of the vector $P(t)$ ($\pi_i \notin S(t)$) are still unchanged.

Step 4: update the the vector probability $P(t)$ as follows:

$$p_i(t + 1) = p_i(t)K(t) \text{ for all i such that } \pi_i \in S(t). \tag{129}$$

- Return to step 1.

Significant developments in both theory and application of learning systems have been made. The field of application of this control approach dealing with complex systems involving large uncertainties widens every day.

9. CONCLUSION

In this part of the book we have presented the most attractive methods for process control purposes. These adaptive control approaches are very simple to implement and need little prior knowledge of the plant to be controlled. The regularized pole placement , the Linear Quadratic Gaussian and long-range predictive control algorithms have shown to be robust with respect to the ubiquitous plant model mismatch and have been successfully applied in the chemical industry (Maurath, 1985; Åström, 1987). The use of learning automata seems to us to be very promising. This control technique is essentially connected to artificial intelligence insofar as human knowledge (know-how) of the plant is included in the performance evaluation unit which is one component of the learning system.

3

IDENTIFICATION AND CONTROL OF LIQUID-LIQUID EXTRACTION COLUMNS

1. INTRODUCTION

Use of liquid-liquid extraction for industrial process is gaining increasing attention in several industries such as the pharmaceutical, petrochemical, nuclear and hydrometallurgical industries. This is because the technique may offer advantages such as energy savings and improved product quality. Therefore, it has become interesting to derive suitable control strategies for such processes.

156

The control objective is to maintain the column at the optimal operating conditions. These conditions were previously defined from hydrodynamic and mass transfer considerations and have resulted in controlling the "hold-up" by acting on:

- pulse frequency for pulsed columns

- rotation speed of the central shaft for Kühni columns

The column considered here was interfaced with an Apple II microcomputer: addressing capacity 64 K, memory (RAM) 48 K, data interface consisting of 16 channel multiplexed successive approximation 12 bit A/D converter, conversion time 25 micro seconds, 4 channel 12 bit D/A converter and a clock. The configuration also involved two floppy disk drives of 140 K of storage capacity each, consol terminale and teletype printer. The host program used in identification and control was written in Pascal. It called external assembly language routines related to real time control: data acquisition routine, digital-analog converter routine and the clock reading routine.

2.   PULSED COLUMN DESCRIPTION

The considered extraction column was a pulsed perforated-plate column. It was 50 mm in diameter and 1 m in height and has 22 perforated trays. The ternary system employed is water-acetone-toluene (system recommended by the European Federation of Chemical Engineering). The solvent (in our case) is the light phase and is dispersed through a perforated pipe

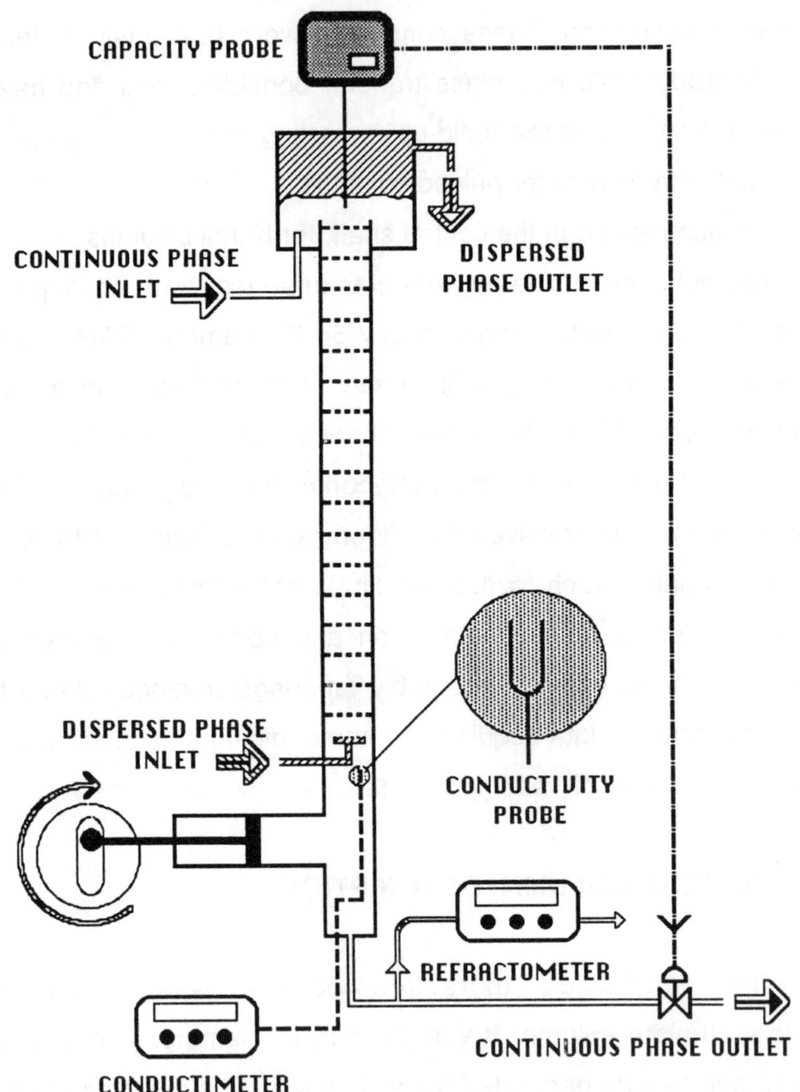

Figure 1. Diagram of the pulsed column

distributor at the bottom of the column. Drops of dispersed phase rise through the continuous phase (water+remaining acetone) to a phase disengaging section. The continuous phase (which is also the heavy phase) is fed in at the top of the column to provide a countercurrent flow. The interface level in the settling zone (at the top) is controlled by means of a capacity probe. A schematic sketch of the pulse column is presented as figure 1. The flow rates are measured by means of electromagnetic flow-meters. A conductivity electrode is set below the light phase distributor and connected to a conductivity meter: this device provides a determination of the amount of dispersed organic phase contained in the aqueous continuous phase. This parameter can also be correlated to the difference of pressures measured over and under the layer appearing below the distributor. This pressure can be easily measured with a differential manometer. The concentration of solute in the raffinate aqueous phase is measured on-line by a differential refractometer. This approach is usually used when the manipulated phases are dangerous and can lead to explosion. The flow rates and the pulse frequency are controlled by a D.C. motor. The amplitude of pulsation has been fixed at 1.5 cm.

The conductivity measured under the distributor and the pulser frequency are selected as control variables.

3. IDENTIFICATION OF A PULSED COLUMN

A single input-output linear discrete model with time varying parameters was adopted to model the complex dynamics of the column.

$$A(q^{-1})y(t) = B(q^{-1})u(t - k) + e(t) \qquad (1)$$

The preliminary studies were made to determine a suitable model order

$(n_a = \text{Degree}\{A(q^{-1})\}; n_b = \text{degree } B(q^{-1})\}$, time delay $\{k\}$, and an appropriate sample rate $\{T\}$ for control (Najim et al., 1986a).

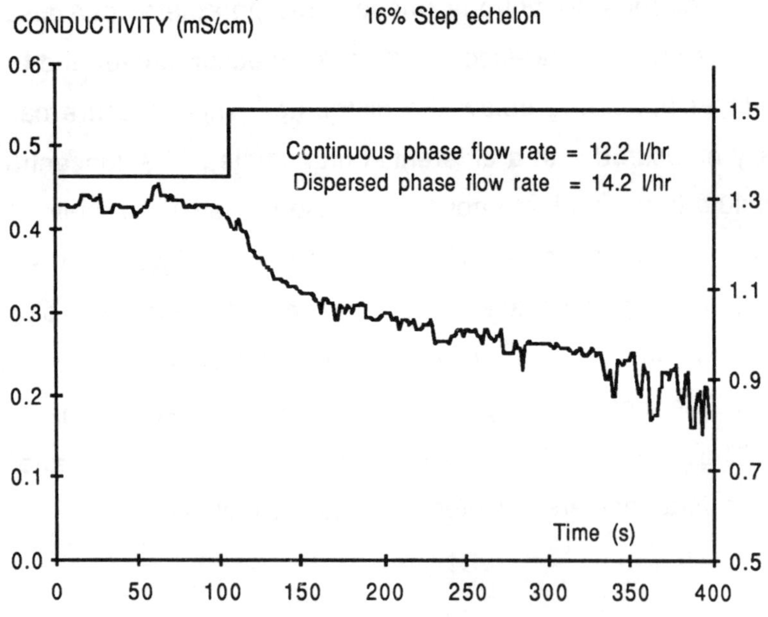

Figure 2. Step response of the pulsed liquid-liquid extraction column

An indication of column dynamics is given in figure 2., which shows the conductivity variation following a change in pulser frequency from 1.35 Hz up to 1.5 Hz.

This step function of about 16% of the frequency has provided a variation of 55% of the conductivity value and may cause a behaviour change.

The values obtained are: $T = 10s$; $n_a = 2$; $n_b = 2$ and $k = 1$.

The process paramaters (coefficients of the polynomial $A(q^{-1})$ and $B(q^{-1})$) are identified by the normalization and projection algorithm developed by L. Praly (1984).The initial values of the gain matrix $F(.)$ and the paramater vector $\theta(.)$ were:

$F(0) = 100I$ (I: the unit matrix)

$\theta(0) = [0.1, 0.1,...., 0.1]^T$

The paramaters associated with the identification algorithm were chosen as follows:

$Mu = 0.7$; $\beta_3 = 5.0$; $\beta_4 = 0.01$; $r_0 = 0.1$; $R = 10.0$; $r(0) = 0.1$; $\beta(.) = 1.0$

To avoid high-frequency variations, low-pass filtered values of the input and output signals were used instead of the actual measurements which were very disturbed. Figures 3, 4, 5 and 6 indicate the variations of estimated model parameters

$$(A(q^{-1}) = 1 + a_1 q^{-1} + a_2 q^{-2} ; B'(q^{-1}) = b_1 + b_2 q^{-1}).$$

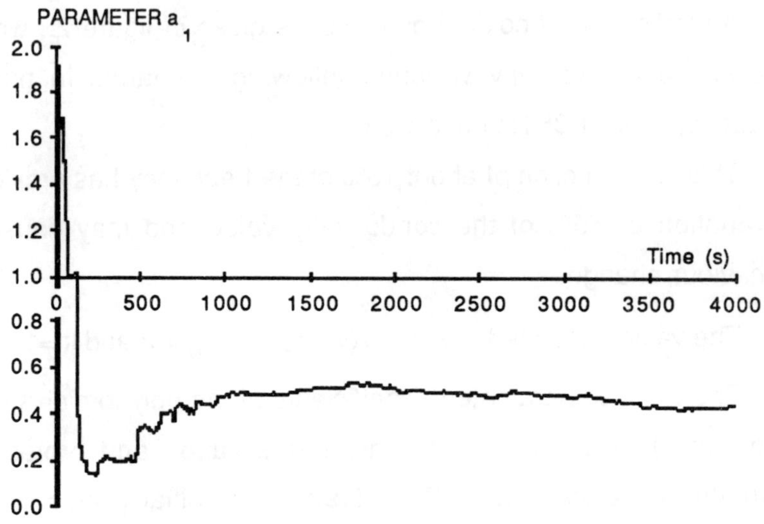

Figure 3. Time evolution of parameter a_1

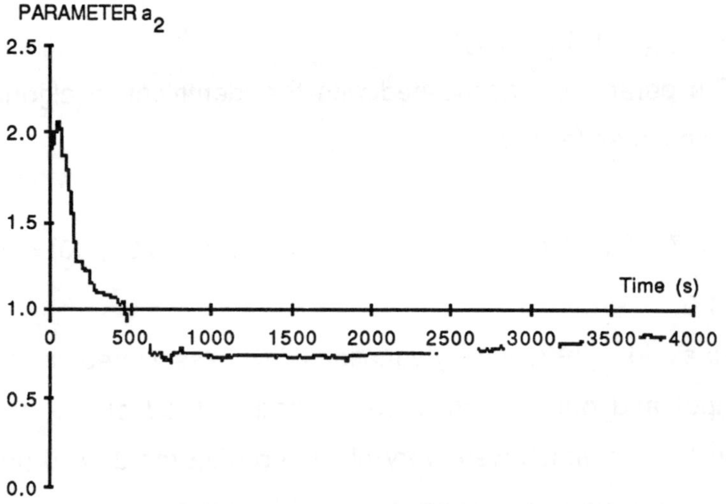

Figure 4. Time evolution of parameter a_2

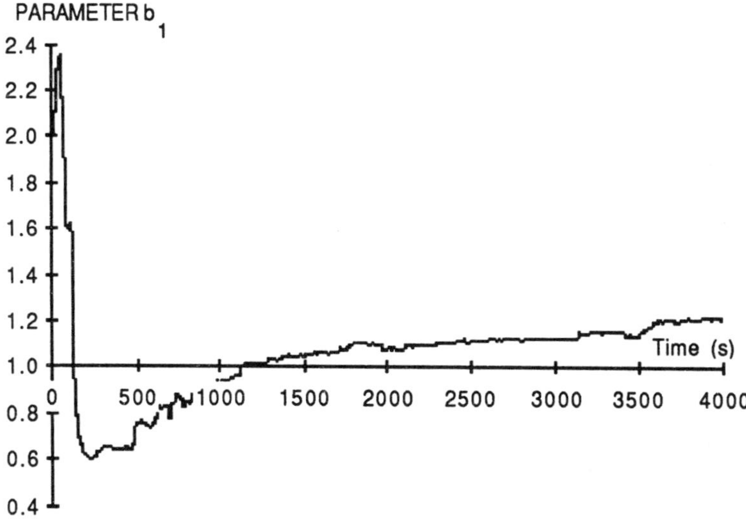

Figure 5. Time evolution of parameter b_1

The dynamics of the column is time varying: the considered system is at first a minimum phase system (up to 3600 s), and becomes a non-minimum phase system (unstable invertible system) afterwards.

Due to round-off, numerical instability may occur after a long time (~10000 sampling periods). To avoid this problem, it is necessary to replace the updating of the gain matrix F(.) by the updating of a factor of F(.) (U-D or square root factorization methods) such that the resulting matrix is guaranted to be non-negative definite (Bierman, 1977; Ljung and Söderström,

1983).

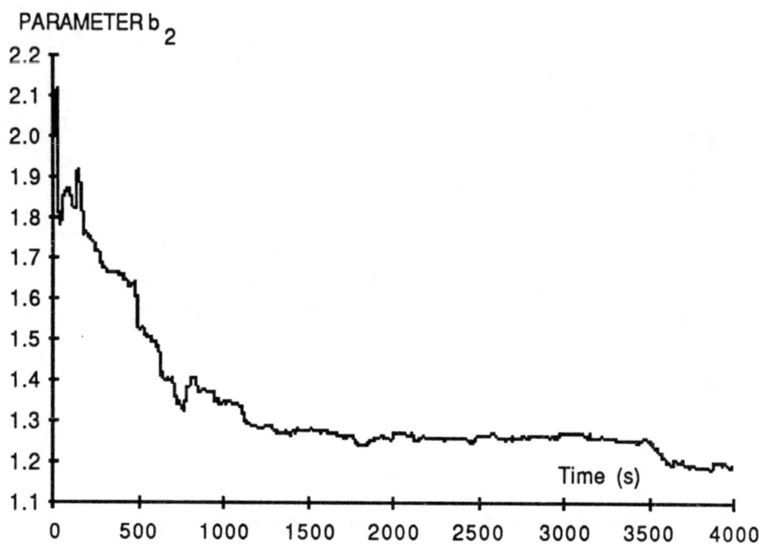

Figure 6. Time evolution of parameter b_2

To obtain an idea of the norm of the parameters, the identification procedure can be initialized by other estimation techniques like the classic least squares method .

4. LINEAR QUADRATIC CONTROL OF A PULSED COLUMN

The application of a linear quadratic self-tuning control approach with restricted domain stability to a pulsed liquid-liquid extraction column is described here. The effective control of extraction columns is a problem to which little effort has been devoted in the past. The application of a self-tuning L.Q.G. with auto-placing of the closed-loop poles in a domain included in the unit circle was, in this instance more a test of the algorithm mentioned above, rather than as a complete assessment of liquid-liquid extraction column control systems.

The considered plant is depicted in figure 1. The control variables are the conductivity measured below the distributor and the pulse frequency (agitation intensity).

We tried unsuccesfully to regulate the considered column using a P.I.D. regulator whose parameters had been adjusted using the Ziegler-Nichols procedure (1943). The dynamics of the column was time varying and very complex. Five types of phase-dispersion behaviours of this column could be observed:

- 1. mixer-settler type operation
- 2. emulsion type operation
- 3. cyclic flooding
- 4. flooding due to emulsification
- 5. flooding due to unsifficient pulsation

This explains that, contrary to other separation techniques such as distillation, few studies deal with the control of extraction columns and the need for adaptive control of such a process. Extraction columns become highly sensitive in the neighbourhood of the optimal operating point., i.e., that which corresponds to the very beginning of flooding. Furthermore, the conductivity is affected by high level noise signal partly due to the mechanical agitation of the pulser.

The implementation of the L.Q.G. self-tuning controller is based on a PDP11 micro-computer using a multitasking microRSX operating system for development of programs and supervision of execution. The main part of the software consists of three concurrent tasks for data acquisition, updating of plant actuators, and for parameter estimation and control algorithm computations. A supervizing program, to control the execution of these tasks, providing in addition other facilities such as plant operation supervision, data logging and real-time adjustment of some critical parameters such as the sampling rate, has also been implemented.

The model parametrization (choice of the input-output polynomial structure and the corresponding polynomial degrees) must be carried out. There are a few parameters that must be specified a priori, i.e., Nr(t) the maximum number of iterations of the Riccati difference equation, the threshold for test of stabilizability (no unstable poles and zeros cancellation) and the desired stability domain (the radius and the coordinate of the center of the circle $C(\delta,\omega)$) and the parameter λ_0.

Using an approximate value of the delay k, a sample interval of 10 s which was selected from the transient response analysis, gave a discrete dead time of 1 sample instant. The control performance is achieved with the following choice of parameters:

- model:

na = 3, nb = 2. The incremental form of the state space representation was used and all initial parameters and signals (y(t) and u(t)) were set to zero.

- controller:

The restricted stability domain corresponds to a circle of radius 0.9 centered at the origin ($\delta = 0.7$, $\omega = 0$).The resetting horizon (Nr(t)) for Riccati difference equation is 10 iterations combined with a test of the blow-up, if any, on the trace of the solution matrix **R**(Nr(t)).

The weighting factor λ_0 was chosen to be equal to 0.2

- identification:

To improve the transient behaviour of the control algorithm, on-line estimation of

parameters was carried out during the start-up phase. This phase takes about 30

samples during which a PRBS (Pseudo Random Binary Signal) control signal is applied to the column under consideration. The control algorithm is then initialized with these parameter estimations.

The parameters associated with the normalization procedure which deals with neglected dynamics are: $\mu = 0.7$ and $r_0 = 0.1$.

A number of experimental results were performed to illustrate the performance of the control algorithm mentioned above.

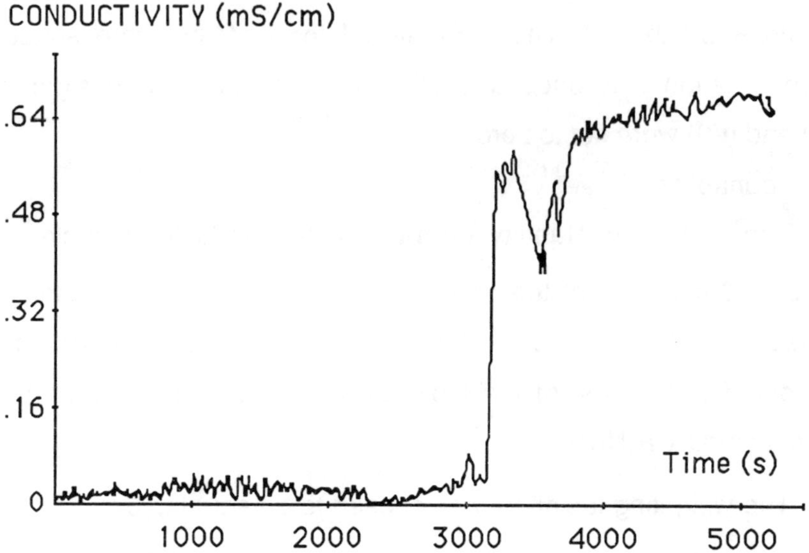

Figure 7. Step response of the column following a negative step
variation of the pulser frequency

Experiments were carried out with the aim of controlling the
column around the optimal operating point corresponding to a
conductivity measurement of about 0.45 mS/cm. The column is
highly sensitive in this region, as can be seen from the step
response of the column for a negative step variation of the pulser
frequency of magnitude 0.3 Hz around the frequency medium
range 1.33 Hz. This step response is presented in figure 7. The
measurements are highly disturbed, showing that filtering of
signals is particularly useful. In the experiments described below

a low pass digital filter was used.

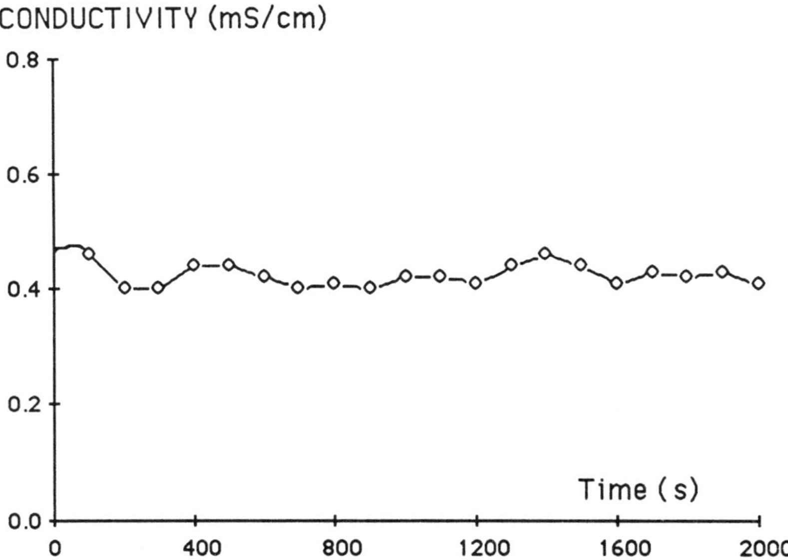

Figure 8. Conductivity variation using self-tuning L.Q.G.

In figure 8 the time dependence of the conductivity has been depicted. The algorithm performs quite well. The conductivity is close to the desired value which corresponds to optimal behaviour of the column. In practice, excessive response to the manipulated variable is undesirable. Figure 9 shows the control action variation. This signal has no energic variations which corresponds to industrial control behaviour requirements. The small fluctuations of the pulser frequency are essentially due to the inconstant dispersed and continuous phase flow rates which are a consequence of a load change. They are time varying. These variations are manually performed by varying the voltage of the D.C. motors which entrain the sucker pumps that feed the columns.

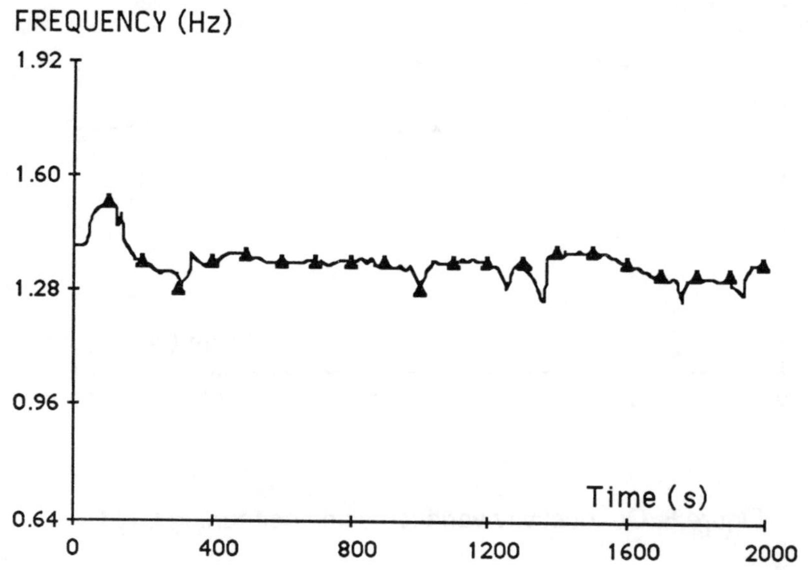

Figure 9. Pulser frequency profile under adaptive control

The dispersed and continuous phase flow rates are respectively depicted in figures 10 and 11. These flow rates are not controlled to keep the column load constant.

The controller and the column model parameters are depicted respectively in figures 12 and 13. In spite of the time variations of the process dynamics, the controller parameter does not vary

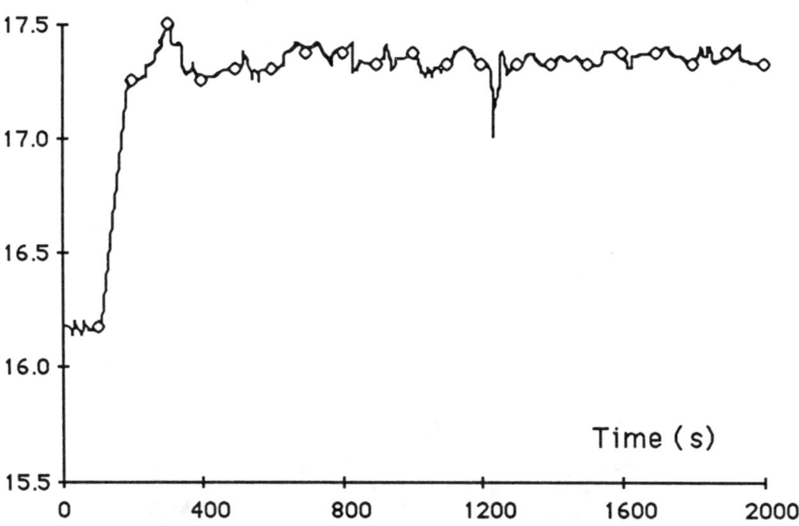

Figure 10. Dispersed phase flow rate profile

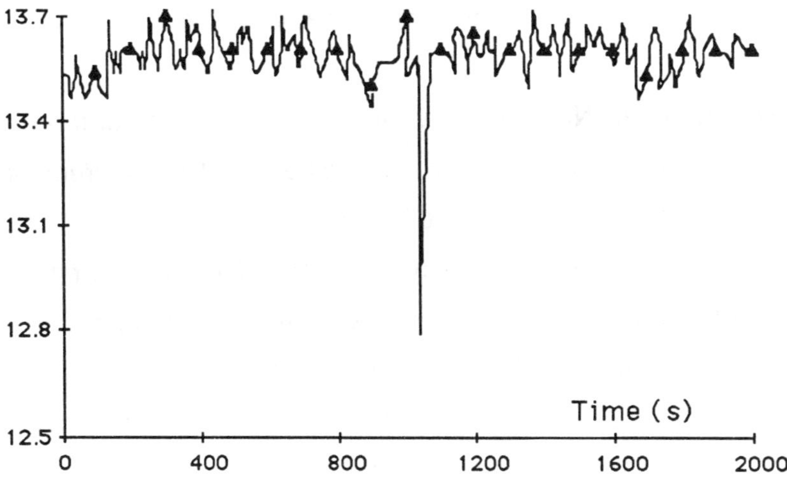

Figure 11. Continuous phase flow rate profile

suddently. The variations are smooth and the controller perfectly follows the variations of the column dynamics.

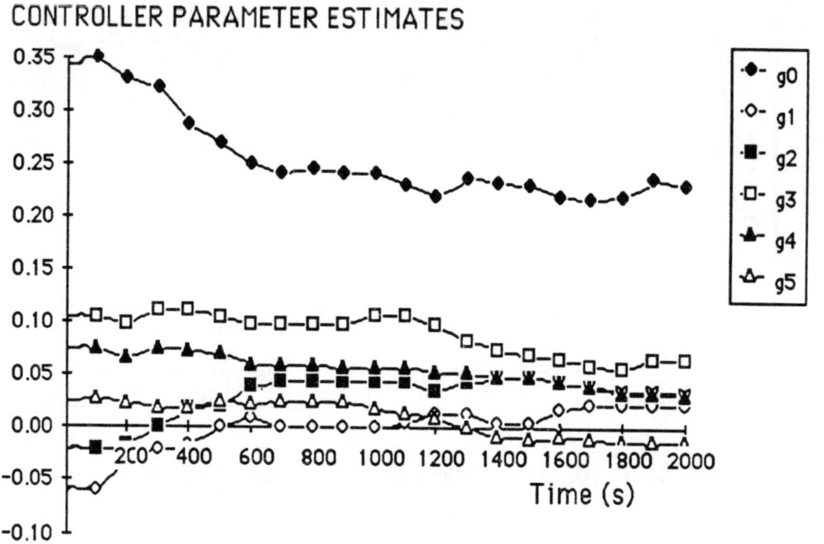

Figure 12. Controller parameter behavior

The number Nr(t) of iterations of the Riccati difference equation needed to fulfil the restricted stability requirements { C(0.7 , 0) } did not exceed three iterations.

The variation of the trace of the matrix **R**(Nr(t)) during the experiments presented is given in figure 14. This trace give some insight into the stabilizability of the closed loop system.

The experiments concerning the implementation of a robust L.Q.G. controller have been presented. This algorithm has potential applications in the control of chemical plants, i.e.,

PLANT PARAMETER ESTIMATES

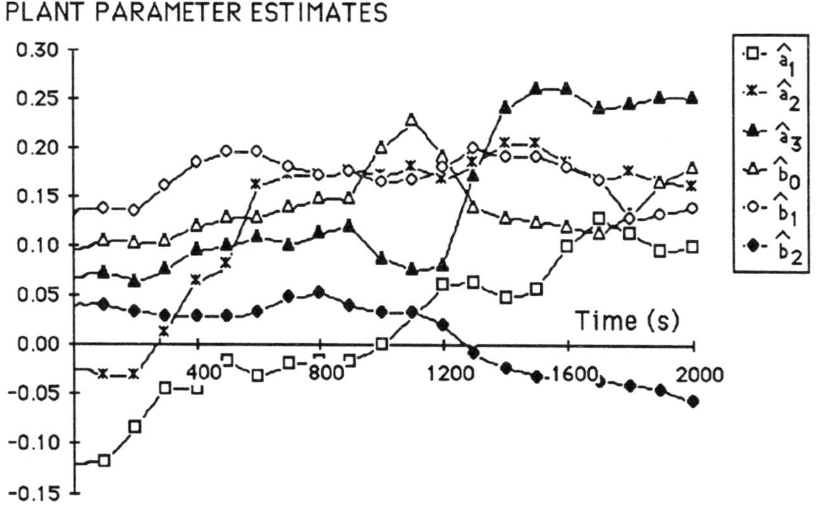

Figure 13. Plant model parameter behavior

MATRIX TRACE

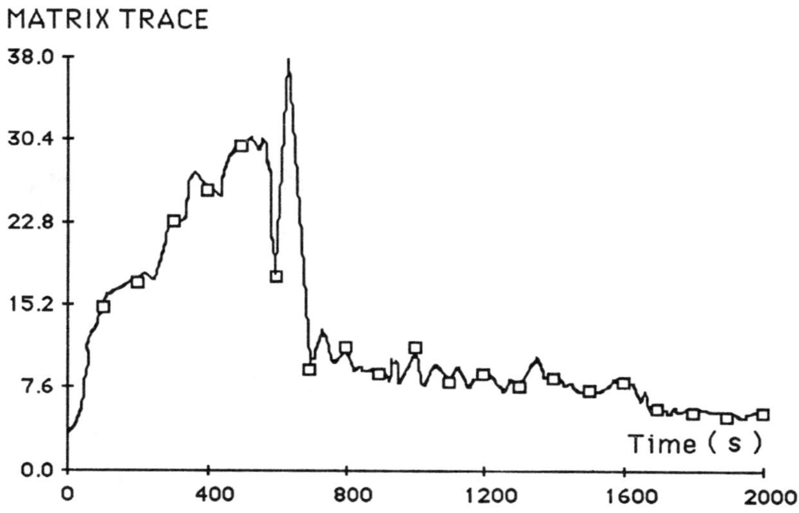

Figure 14. Variation of the trace of the matrix $\mathbf{R}(Nr(t))$

processes which are often non linear, time varying and subject to several kinds of disturbances. Inverse responses (non-minimum phase systems) and open-loop instability are generally observed in these processes.

The L.G.Q. controller with restricted stability domain proved to be effective in ensuring closed-loop stability and satisfactory dynamics response which can be specified a priori by choosing the values of the parameters δ and ω. These specifications can be easily made in the frequency domain.

In the following we shall be concerned with the implementation of the long range predictive control algorithm.

5. PREDICTIVE CONTROL OF A PULSED COLUMN

This paragraph will be present the application of the long-range predictive control algorithm to the pulsed liquid-liquid extraction column (Najim et al. 1986b).

The control objective was to maintain the conductivity measured under the distributor, around the desired value: about 0.45 mS/cm, which corresponds to a mean value between the conductivity of water and toluene respectively. Initially the process parameters were estimated by decreasing gain with a forgetting factor. To avoid parameter drift, the algorithm proposed by L. Praly (1984), and presented in chapter IV, is then applied for computing the process paramaters. The obtained results (figures 5 and 6) show that in transient behaviour, the system becomes inverse unstable. The output and the control horizon are Nu = 1; Ny = 6. The experimental results allow automatic starting of the column (the dispersed phase flow rate was increased from zero to about 16.0l/Hr) and remains excellent when the column behaviour changes.

The characteristic polynomial $P(q^{-1})$ was chosen to be equal to $(1 - 0.8q^{-1})^2$.

The influence of the weighting factor λ is negligeable. It was chosen to be equal to 0.01.

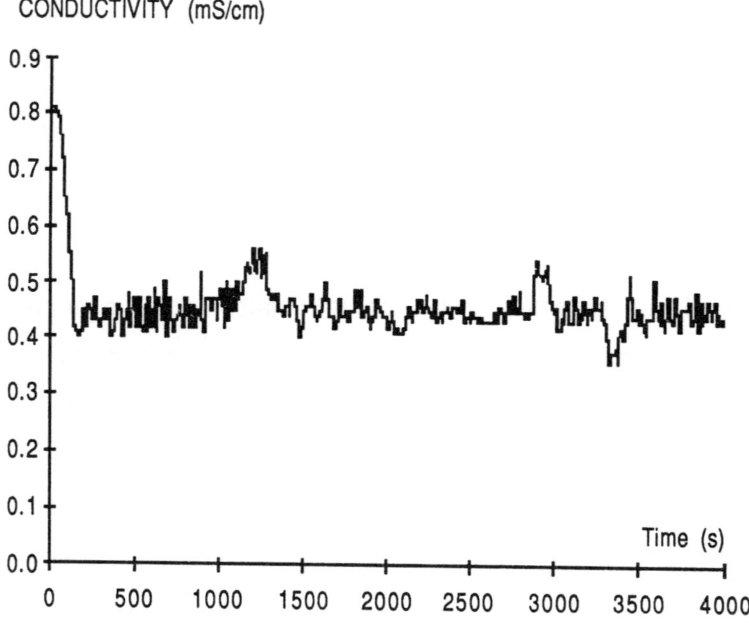

Figure 15. Time evolution of conductivity

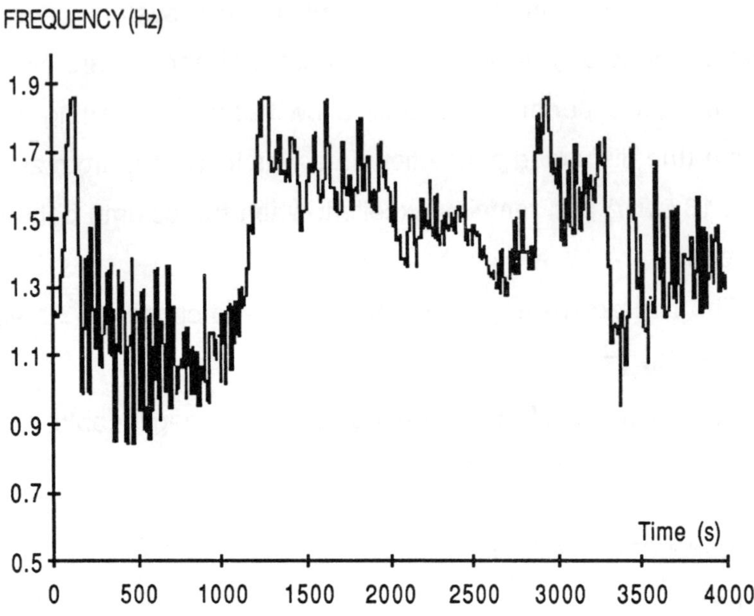

Figure 16. Time evolution of pulse frequency

In spite of highly disturbed flow rates which considerably modified the dynamics of the column, this algorithm brings the column to and maintains it at the optimal operating conditions (product quality, regular behaviour, energy saving, etc.). The starting up horizon time is less than 2 minutes. Figures 15, 16, 17 and 18, show the conductivity variations, the pulse frequency, the continuous phase and the dispersed phase flow rates respectively.

In figure 15, the evolution of the conductivity allows analysis of the controller's behaviour. At the initial time, the conductivity is

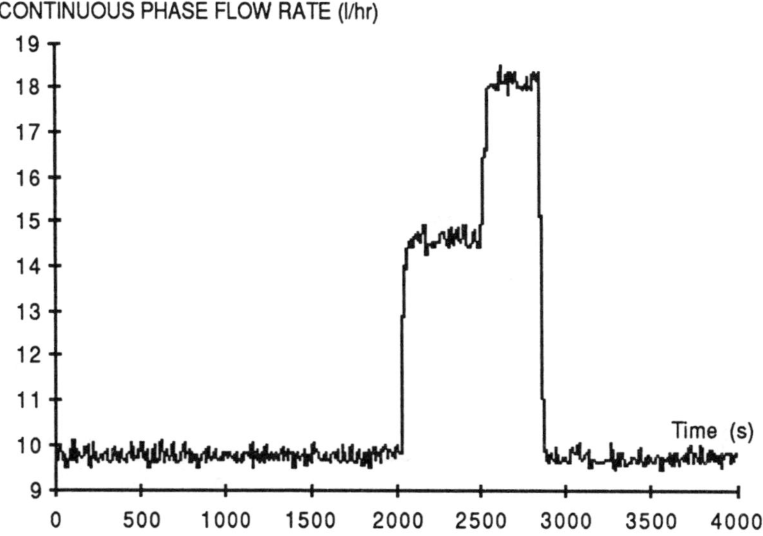

Figure 17. Time evolution of continuous phase flow rate (water)

equal to 0.8 mS/cm, which is the value for the water phase alone. Then, while the dispersed phase flow rate is increased from 0 to 16l/Hr within 200s (see figure 18), the controller brings the operating point to flooding point, i.e., to a conductivity around 0.45 mS/cm. Afterwards, in spite of important step changes of both flow rates, the controlled variable remains almost unchanged, which means that flooding is just maintained.

Moderate peaks come with step changes in the flow rates : for example, at time   t = 1000 s, a decrease of the dispersed flow rate causes a slight increase of the conductivity, which means a certain underflooding of the column, but quickly, the controller increases the intensity of pulsation (see figure 15) so that flooding is

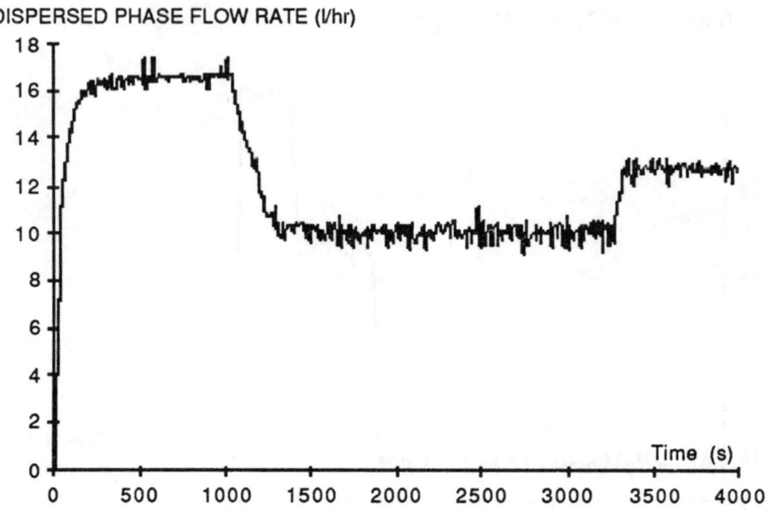

Figure 18. Time evolution of dispersed phase flow rate (toluene)

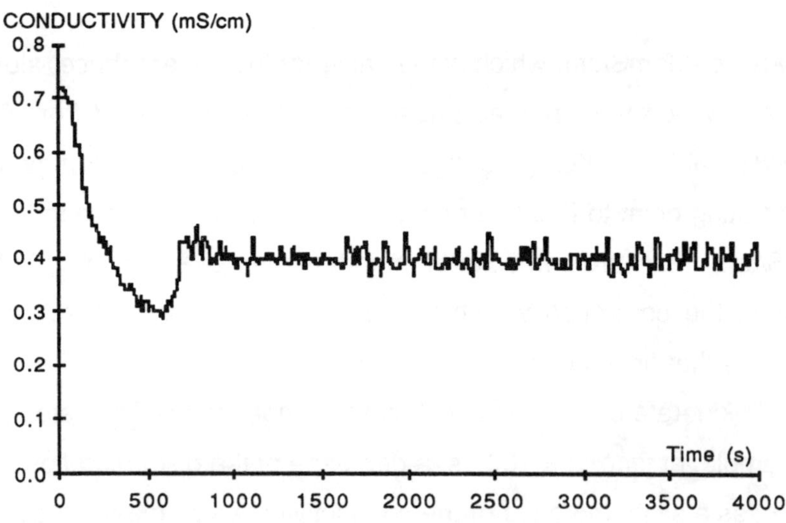

Figure 19. Time evolution of conductivity

recovered. Similar behaviour can be noticed with step changes on the continuous phase flow rate at time $t = 2000$ s, $t = 2\,500$ s and $t = 2800$ s; the controller seems to be less affected by these strong perturbations.

The following experiments were carried out with an aqueous solution of 5 weight % of acetone as continuous phase. The dispersed phase was composed by toluene. In this case we were concerned with mass transfer (transfer of acetone from the continuous to the dispersed phase). Figure 19 shows the conductivity variation (the desired value was 0.4 mS/cm).

With mass transfer, the dynamics of the column has changed and, for example, during the start-up period, one can notice in figure 19 that the column gets slightly over-flooded. This is due to the fact that in order to flood the column rapidly , the controller tends, at first, to greatly increase the intensity of pulsation (see figure 20), which causes an intensive breakage of drops and the formation of small droplets.

Because of mass transfer, coalescence between drops is hindered and even when the intensity of pulsation decreases, the diameter of drops cannot grow as rapidly as required, and the column suffers from a transient light overflooding. Nevertheless, in spite of changes in the dynamic behaviour of the column due to mass transfer, the controller proves to be able to adapt itself.

Figures 20-21 and 22 show the variation of pulse frequency,

continuous phase and dispersed phase flow rates respectively. The variation of the pulse frequency is smooth, consequently the actuator is not drastically called upon. This behaviour constitutes another source of economy; because it leads to long conservation of the material.

The variation of the continuous and dispersed phase flow rates are very important. At certain times, these variations are great erthan 10%. These variations have been voluntarily imposed on the column to simulate industrial behaviour where flow rates are generally time varying.

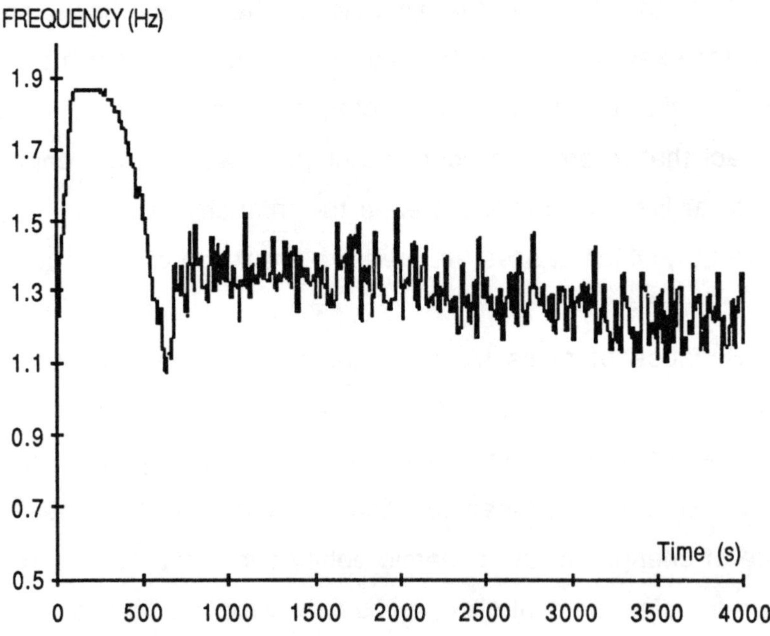

Figure 20. Time evolution of pulse frequency

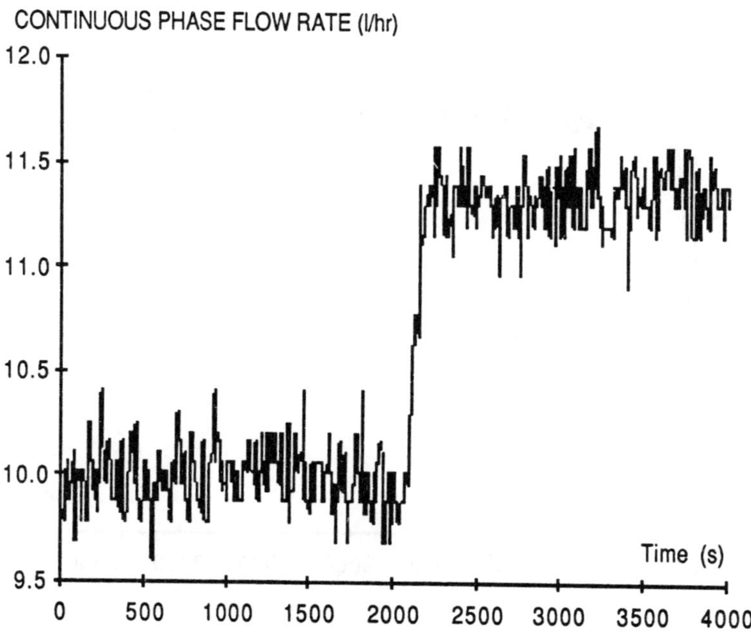

Figure 21. Time evolution of continuous phase flow rate
(water+acetone)

The great change between operations respectively with or without mass transfer shows that the operating point under steady-state conditions as well as the dynamic response of the column depend strongly on the mass transfer rate, this dependency being strictly impossible to predict and to modelize. This emphasizes the efficiency of the implemented control algorithm to adapt itself to time-varying conditions.

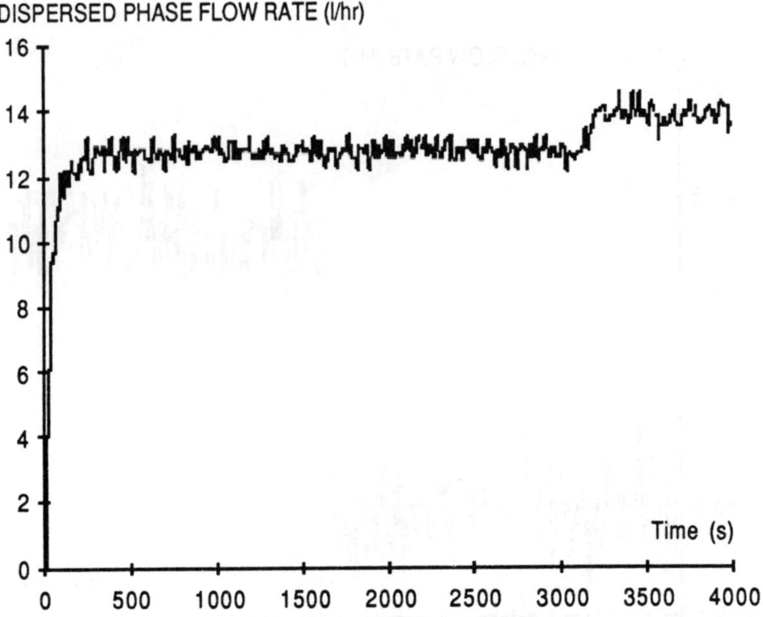

Figure 22. Time evolution of dispersed phase flow rate (toluene)

The evolution of acetone concentration in the continuous phase (water + remaining acetone) outlet is depicted in figure 23. The response of the column in terms of acetone concentration in the continuous phase is slow, compared to the column response following a pulse frequency agitation. Nevertheless the dynamical behaviour of the column under adaptive predictive control algorithm is good. In fact, after a transient time, the concentration of acetone in the continuous phase is close to the desired value.

In order to deal with a stabilizable model (with no common unstable zeros and poles), one solution consists of simplifying common unstable poles and zeros of the estimated process model. As the identication procedures are mapping the data into a certain order model set, model reduction can be performed to

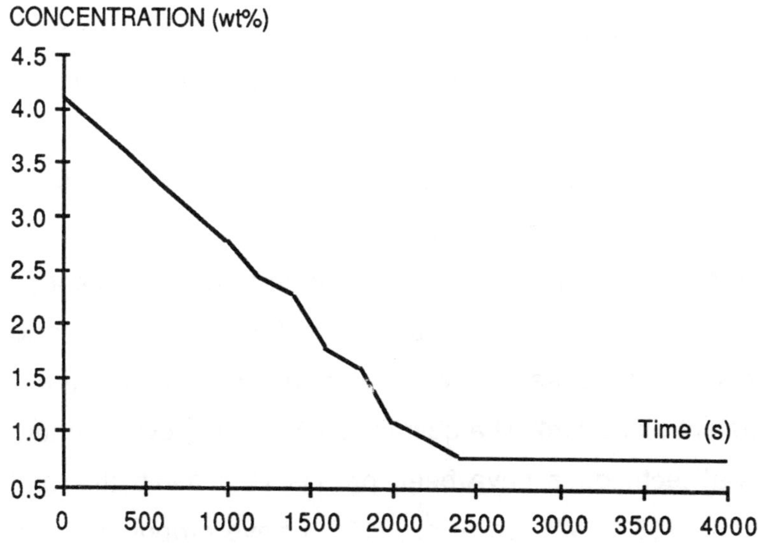

Figure 23. Time evolution of the concentration of acetone in continuous phase outlet

reduce the order of this high order estimate so as to obtain a lower order model which may be stabilizable. Praly et al. (1988) have exercised a direct adaptive control on pulsed liquid-liquid extraction column which deals with nonstabilizable systems. In this study, the control algorithm is derived using the pole placement method as the underlying design procedure and is based on estimation of both model and controller parameters. This estimation procedure is bilinear in these parameters and is obtained from input and output prediction error models. A least sqares criterion minimization is adopted to solve this identification problem. In this approach no assumptions are made about the stabilizability of the identified model nor about the persistency of excitation nor about the knowledge of an upper bound of the parameters and of the system order. Studies of robustness towards the neglected dynamics and towards bounded perturbations were carried out by Praly (1984).

When an adaptive control algorithm is used in steady-state regulation of near deterministic systems (the common behaviour of chemical processes), the data gives little information (non persistent excitation). The gain matrix F(.) will grow exponentially. Several techniques have been proposed to avoid this problem (Åström, 1983). M'Saad et al., (1986) have proposed a method which consists of keeping the estimate of parameters within a certain predefined domain in the case of poor excitation; i.e., when an information measure is less than a certain bound. The introduced information measure is:

$$m(t) = X^T(t - 1)F(t - 1)X(t - 1) \qquad (2)$$

m(t) measures how much the recent information differs from the past data. It always lies in the range $0 \leq m(t) \leq 1$.

The control studies reported here demonstrate a successful application of self-tuning, based on generalized predictive control with multiple model reference, to a pulsed liquid-liquid extraction column. Some previous knowledge is necessary to implement this algorithm. The performance of this algorithm is not sensitive to the choice of model parameters (order, time delay, sampling rate, etc.).

6. LEARNING CONTROL OF A PULSED COLUMN

The characteristics of swarms may be analysed in terms of drop transport due to the Archimedes force, and breakage and coalescence due to mixing intensity. Most of these mechanisms are due to turbulence which is developed in the column by means of the pulsing device: as a matter of fact, drop breakage may be viewed as a random mechanism caused by turbulence, and coalescence may be analysed as the consequence of random contacts between drops in this agitated medium.

The control problem is modeled as that of a learning automaton operating in a random environment (Najim, 1982). The environment models the controlled process (pulsed column). The column is connected in a feedback loop to a stochastic automaton which considers the controlled process as a random medium. The learning system collects and processes data (input-output measurements) and changes its internal structure and the associated probability distribution to improve (in some specified sense) the performance of the column. The learning system consists of a performance evaluation unit and a stochastic automaton with variable structure.

The control objective is to maintain the conductivity measured under the ditributor close to a desired value corresponding to the appearance of flooding (about 0.45 mS/cm). Know-how on the process is introduced in the performance evaluation unit. In the present case, this know-how has been reduced to its minimum and consists of the following rules:

 - if the conductivity measure is above the desired value, the pulse frequency must be increased to reach flooding conditions.

 - if the conductivity measure is below the desired value, the column is overflooded and the pulse frequency must be decreased.

The system performance generates either reward or penalty according to this control objective which is formulated as follows:

Let $y(t)$ be the process output (the conductivity of the medium measured below the distributor).

<u>$w(t) = 0$</u>

If $\{ y(t) <$ the desired value and $u(t) < u(t-1)\}$

or

$\{ y(t) >$ the desired value and $u(t) > u(t-1)\}$

<u>$w(t) = 1$</u>

otherwise

The optimal reinforcement scheme developed by Seret and Macchi (1982) was employed. That scheme is given as follows:

Let u_i be the action (control) chosen at time t and $p_i(t)$ its associated probability, then:

$\underline{w(t) = 0}$

$$p_i(t + 1) = \{p_i(t) + \beta_0 p_i(t)[1 - p_i(t)]\}/\alpha$$

$$p_j(t + 1) = \{p_j(t) - \beta_0 p_i(t) p_j(t)\}/\alpha \qquad j \neq i \qquad (3)$$

The index j corresponds to variable actions which have not been selected at time t.

The quantity: $\alpha = p_i(t) + .. + p_N(t)$ is introduced to avoid the numerical problems due to round-off errors, and to guarantee that $p_i(t) \in [0,1]$.

$\underline{w(t) = 1}$

The probability associated with the action u_i will be decreased and the others
$(j = 1,N; j \neq i)$ will be increased.

$$p_i(t + 1) = \{p_i(t) - \beta_1 p_i(t)[\ p_i(t)]\}/\alpha$$

$$p_j(t + 1) = \{p_j(t) + \beta_1 p_i(t) p_j(t)\}/\alpha \qquad j \neq i \qquad (4)$$

with $0 < \beta_0 \leq 1$ and $0 \leq \beta_1 \leq 1$

This reinforcement scheme is optimal and is easy to

implement.

The flexibility and the clarity of this formulation are very interesting to represent practical and theoretical knowledge of the process to be controlled.

Initially, the column was brought manually near its optimal operating point. Then the column was swiched over to the microcomputer. The control range was discretized into a set of 20 intervals ($N = 20$; $u_{min} = 0.96$ Hz; $u_{max} = 1.73$ Hz). No prior information related to the process was taken into consideration. The state probability was initialized as follows: $P(0) = [1/N,....,1/N]^T$.

The parameters β_0 and β_1 were chosen to be equal to $\beta_0 = \beta_1 = 0.1$. It is to be noticed that the adaptation of parameters β_0 and β_1 can be done according to change of the operating conditions to improve the tracking capability of this algorithm (by analogy with the identification techniques used for parameter change detection: adaptation of the forgetting factor and the trace of the gain estimator, Favier and Di Martino, 1984).

For the first experiments, the conductivity reference (desired output) was taken to be equal to 0.45 mS/cm. The output variable was measured at time t and the control action was calculated and applied to the column at time $t + \tau$, where τ is the computation time related to the control algorithm defined above ($\tau = 3$ s). The continuous and dispersed phase flow rates are subject to perturbations. They are not controlled. The behaviour of the column under learning control is shown in Figure 24 which represents the conductivity time variation, illustrating the ability of the learning algorithm to adapt itself to the variations of the parameters affecting the behaviour of the column. The conductivity measurements were not filtered before being processed by the algorithm. An introduction of an analog or digital second-order filter may lead to smoother signals and improve the performances of the control system (practically all industrial sensors have some kind of filter).

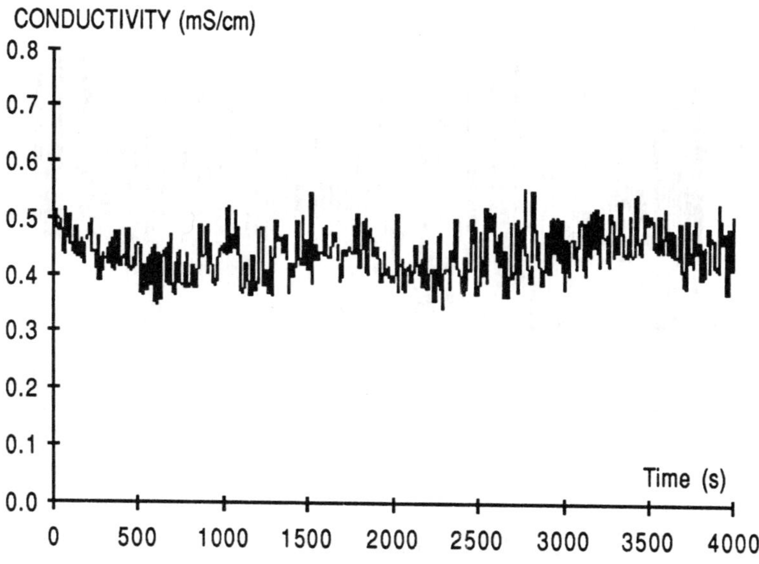

Figure 24. Time evolution of conductivity

Figures 25, 26, 27 indicate the time variations of the pulse frequency, the continuous phase and the dispersed phase flow rate respectively. The dispersed and continuous flow rates which are supplied by pumps are highly disturbed. These pumps are nonlinear characteristics. Experiments carried out off-line show that the filtering of the pulse frequency action using a simple first order low-pass filter leads to a smooth signal.

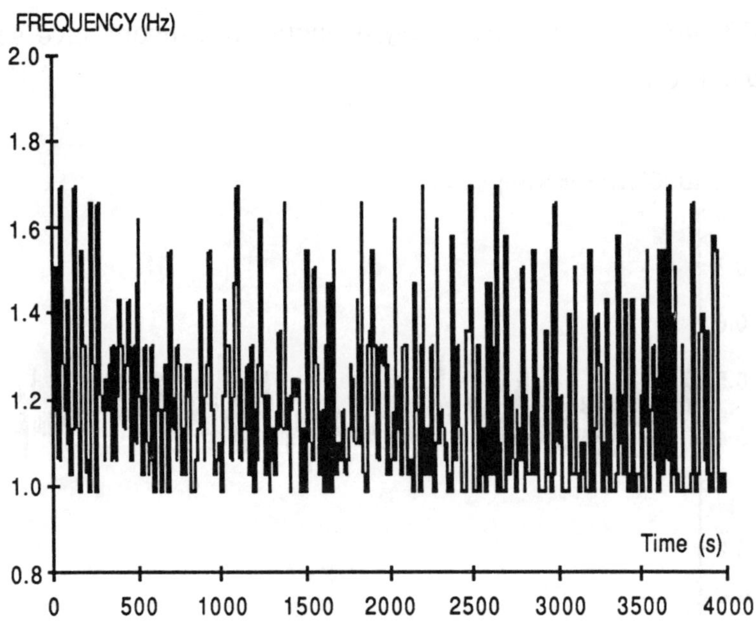

Figure 25. Time evolution of pulse frequency

A P.I.D. regulator was initially used to regulate the conductivity. Based on the Ziegler-Nichols (1943) tuning procedure, the P.I.D. constants were chosen. The performance of this regulator was poor and led to a non-desirable column behaviour (countercurrent flow through the column was no longer occuring and reversal of the continuous phase could happen in some cases, etc.). These results confirmed the necessity of implementing control strategies which parameters are adjusted on-line.

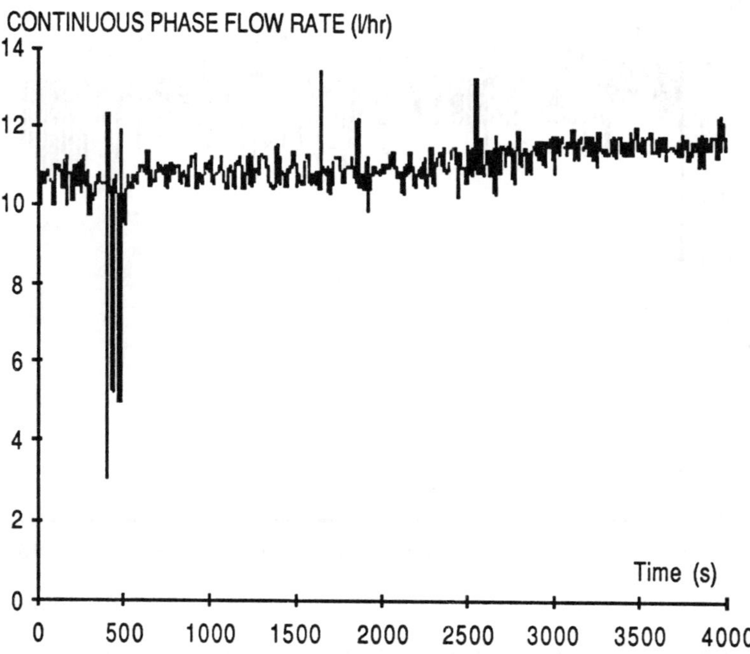

Figure 26. Time evolution of continuous phase flow rate (water)

We can notice also that the rule introduced in the performance evaluation unit are very simple. They can be improved to take correctly into consideration the complex dynamics of the considered column. It can be interesting to introduce other analytical knowldege on the column in this control unit. This knowledge may take the form of models, correlations, etc.

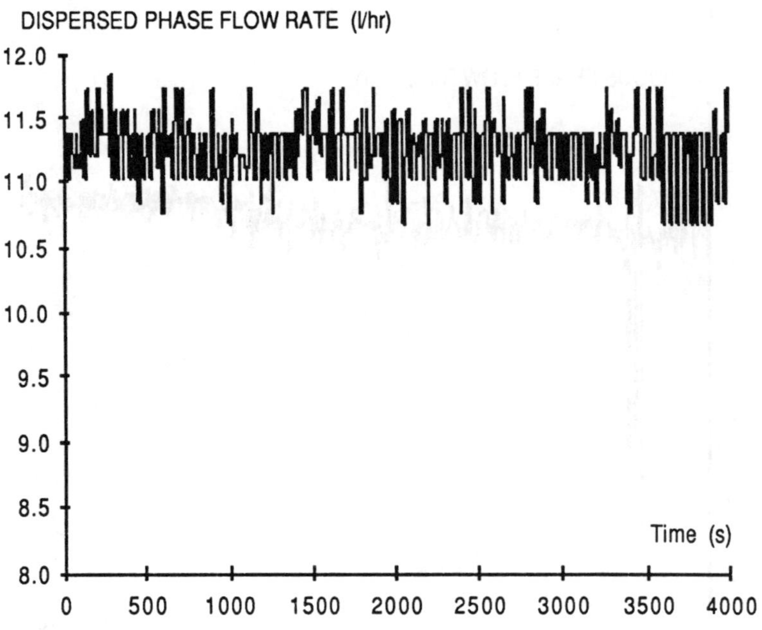

Figure 27. Time evolution of dispersed phase flow rate (toluene)

The state probability distribution is summarized in the following figures. Figure 28 represents the probability attributed to the optimal control action which corresponds to the first control interval. Figure 29 gives the same representation for the least probable control interval (interval n°9). Figure 30 shows the overall state probability distribution at different times. These probabilities give an idea of the dynamics of the column. After a transient period convergence occurs. After 2600 s, the probability associated to the first control action is approximately equal to 0.25. While the probability associated to the control interval numbered 9 is practically null. This means that, during this behaviour period, this control action will not be selected and consequently not applied to the column.

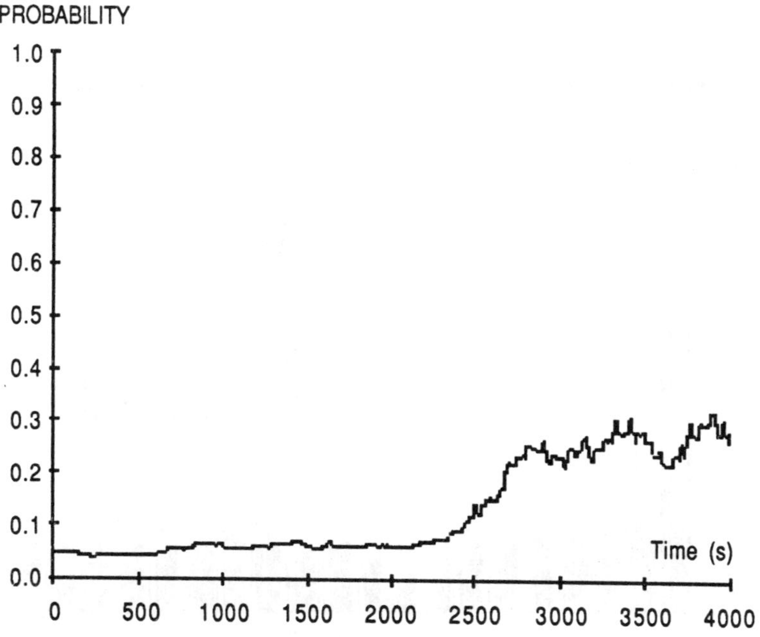

Figure 28. Time updating of the probability for interval n° 1

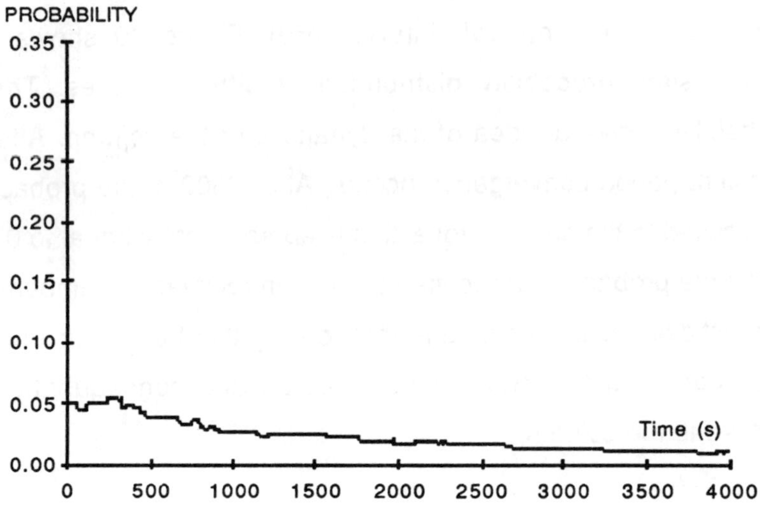

Figure 29. Time updating of the probability for interval n° 9

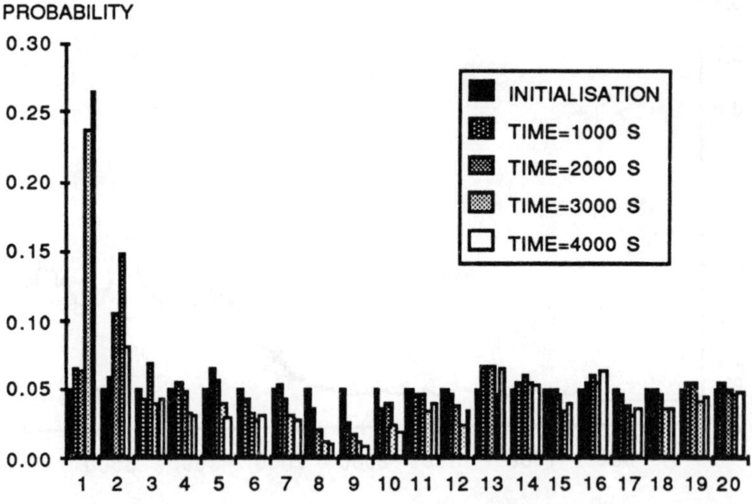

Figure 30. Values of the probabilities for different times and
intervals

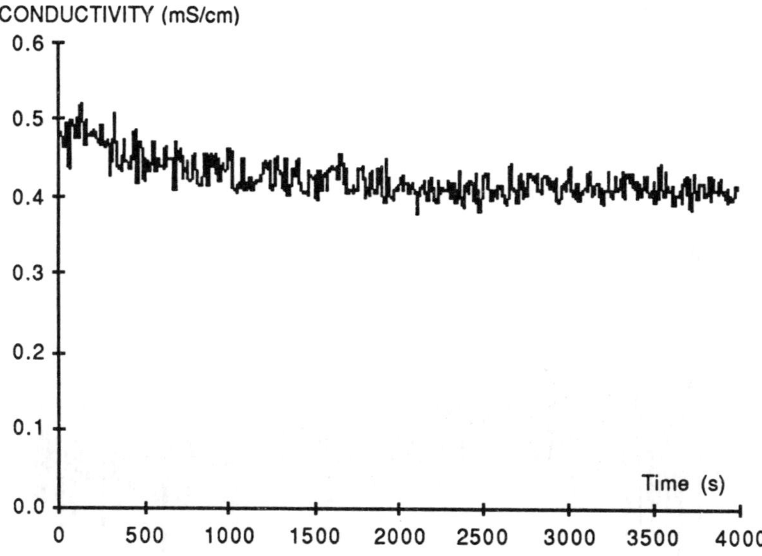

Figure 31. Time evolution of conductivity

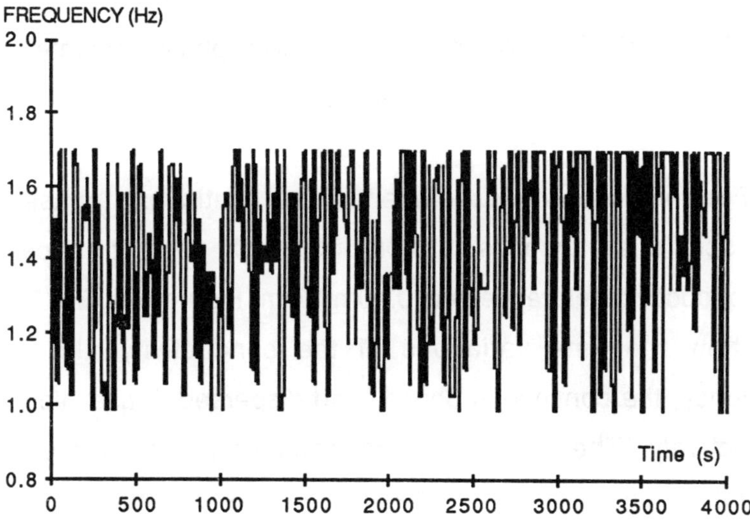

Figure 32. Time evolution of pulse frequency

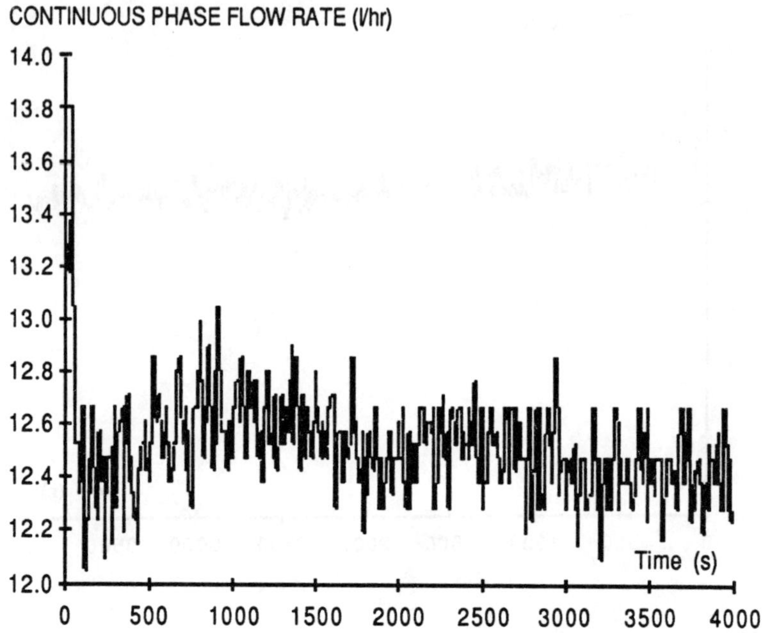

Figure 33. Time evolution of continuous phase flow rate
(water+acetone)

For the following experiments, the continuous phase was composed of an aqueous solution of 5 weight percent of acetone; the conductivity reference was 0.40 mS/cm. Figures 31, 32, 33 and 34 show the time variations of the conductivity, the pulse frequency, the continuous phase and dispersed phase flow rate respectively. The dispersed and continuous flow rates were subject to high variations. These variations disturb the column considerably. In these conditions we can not achieve the desired performance by using a fixed parameter regulator.

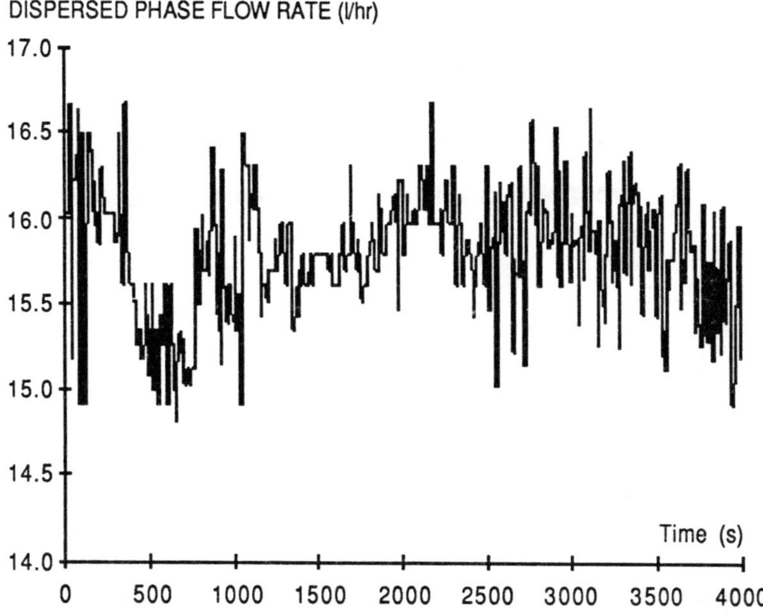

Figure 34. Time evolution of dispersed phase flow rate (toluene)

The presence of acetone changes to a considerable extent the process dynamics by introducing mass transfer mechanisms. These phenomena are very complex and do not occur only in the column. The phenomena of transport and coalescence still exist. All these mechanisms depend on the operating point of the column and on the chemical and physical properties of the considered phases. The fluctuations of both flow rates are quite important to simulate practical use of liquid-liquid extraction column as, for example, in fine chemistry where the plant must be polyvalent and production needs vary from one time to another.

Figure 35 represents the probability evolution attributed to the optimal control action (u_{20} =1.69 Hz) while figure 36 gives the same representation for the least probable one (u_6). Finally, the overall state probability distribution is depicted in figure 37.

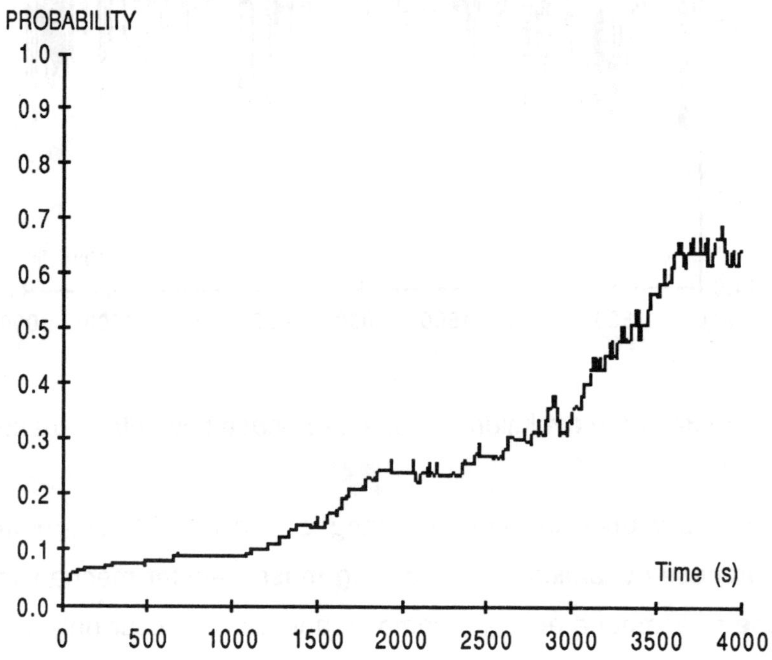

Figure 35. Time updating of the probability for interval n° 20

The probability associated with optimal control action was approximately equal to 0.7, while the initial probability which is associated to it was equal to 1/N where N is the number of control actions. We must recall that the control action does not vary continuously and the number of actions is small, to correctly represent the possible range of the pulse frequency, i.e., the control variable domain variation. An increase of this number of actions may lead to better control performances. Another rule can also improve the performance of the learning control system: it consists of freezing the probability vector, or to apply the same control action (the control selected at time t) when the selected one (the control computed at time t + T, where T is the sampling rate) is close to the control signal previously applied to the column under consideration.

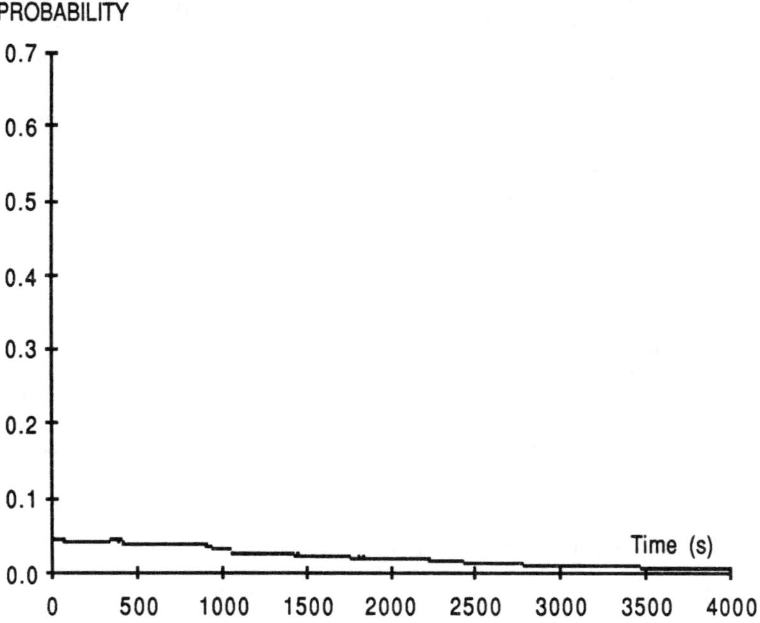

Figure 36. Time updating of the probability for interval n° 6

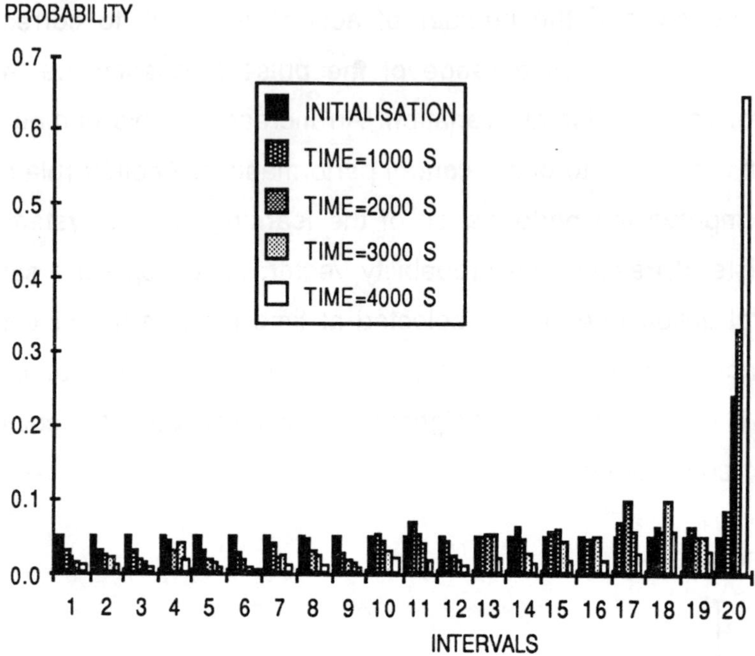

Figure 37. Values of the probabilites at different times

A learning automaton has been used to control a pulsed liquid-liquid extraction column and has shown good control performance (Najim et al., 1987b).

This approach is well suited for industrial applications where it is difficult to develop a mathematical model or where the application of the optimal control theory leads to complex solutions that are unsuitable for real time control.

The algorithm presented is easy to implement as a monitor routine in a process computer system.

In the next section we shall be concerned with the use of

hierarchical system of learning automata which is a solution for reducing the number of actions to be updated and consequently the computer time.

7. MULTILEVEL CONTROL

A pyramidal structure of learning automata with 3 levels was used to control the column (Thathachar and Ramakrishnan, 1981). The numbers of actions for each automaton in the hierarchy is equal to 3 (N = 3). The control domain variation was consequently discretised into 27 intervals. The control algorithm was specified initially with $P(0) = [1/N,......,1/N]^T$; $\mu(.) = 0.05$.

The flow rates are measured and controlled respectively by flow-meters and pumps. The pulse frequency is controlled by a D.C. motor.

The controlled variable is measured at time t and the the pulse frequency is calculated and applied to the D.C. motor which controls its variation at time $t + \tau$, where τ is the computation time related to the pyramidal structure of learning automata which have been implemented on an Apple II microcomputer (τ = 4s). The continuous and dispersed phase flow rates are not controlled.

For the first experiment, the desired value of the conductivity was 0.45 mS/cm (which corresponds to flooding conditions). Figures 38, 39, 40 and 41   present time evolution of the conductivity, the pulse frequency, the continuous phase and dispersed phase flow rate, respectively. These results are better that those obtained with a single automaton. The conductivity measurements have not been filtered and the continuous and dispersed flow rates are subject manually to variations (about 50% at certain times).

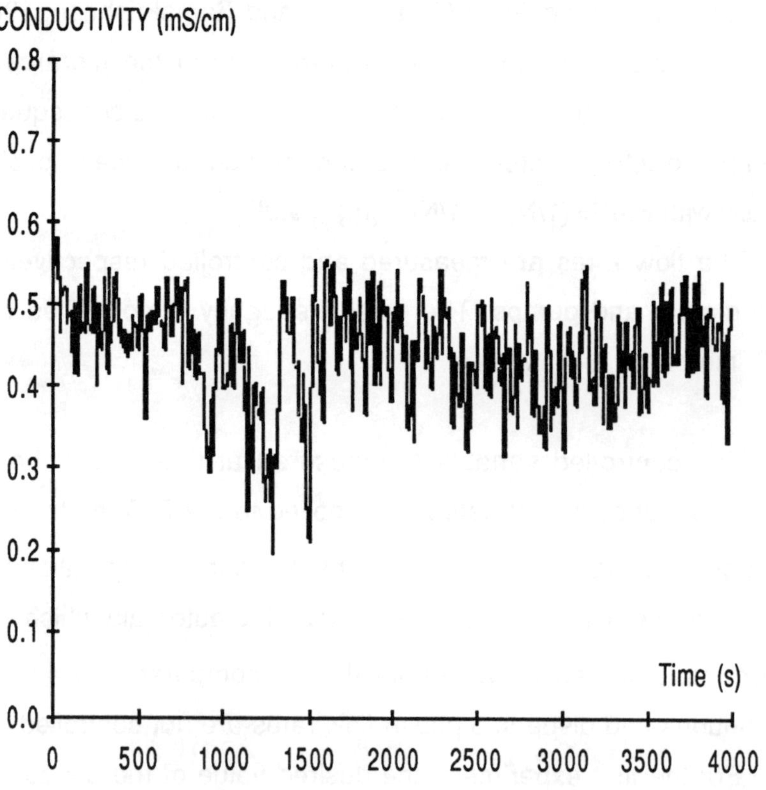

Figure 38. Time evolution of conductivity

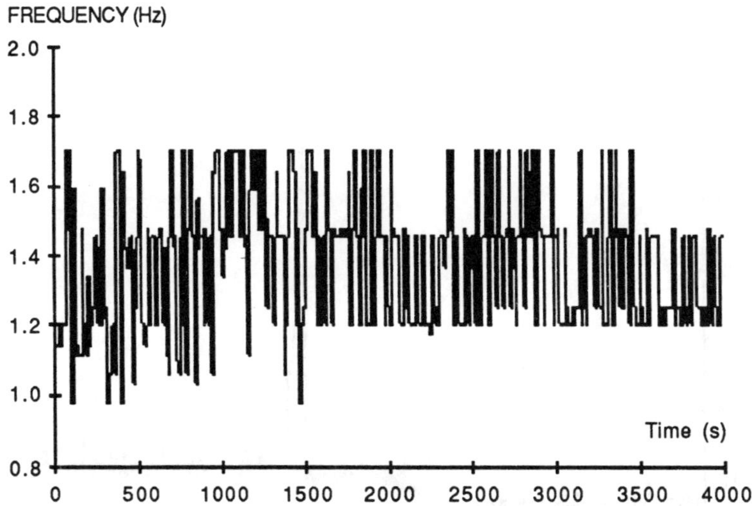

Figure 39. Time evolution of pulse frequency

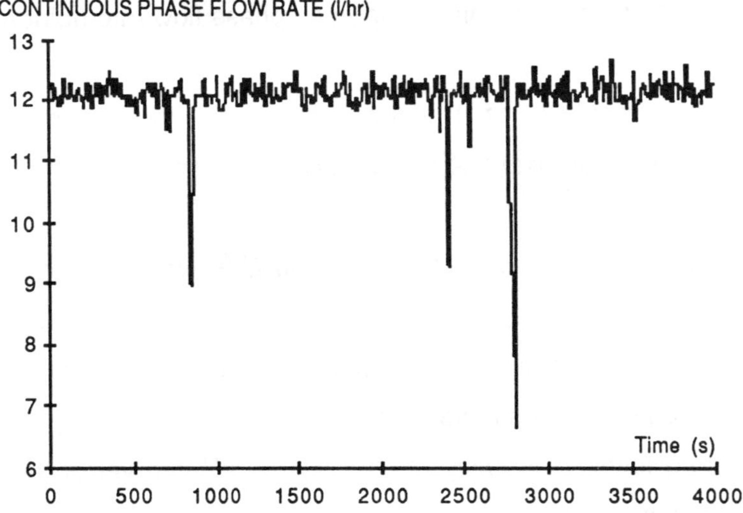

Figure 40. Time evolution of continuous phase flow rate

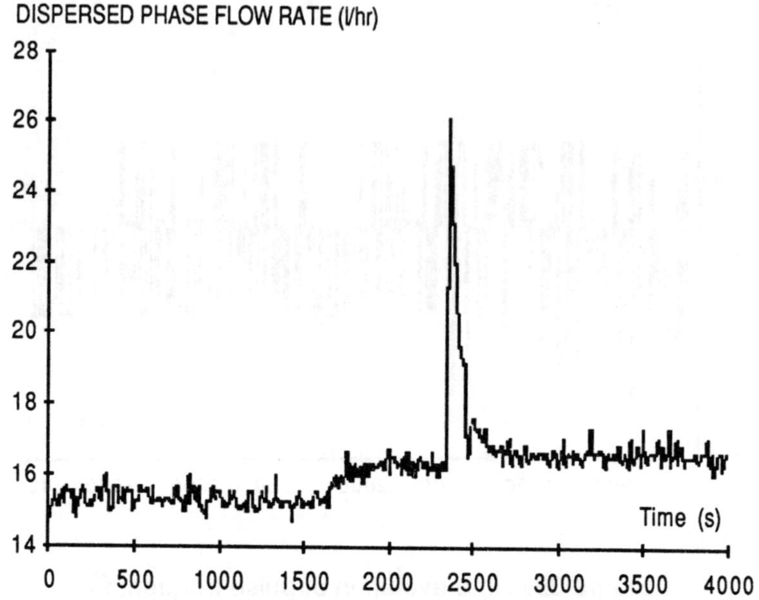

Figure 41. Time evolution of dispersed phase flow rate (toluene)

Time updating of the probability distribution can be summarized by an example of probability evolution for each level:

- first level: probability related to the 2nd interval as in figure 42.

- second level: probability related to the 3rd interval of the 2nd automaton as in figure 43.

- third level: probability related to the 3rd interval of the 6th automaton as in figure 44.

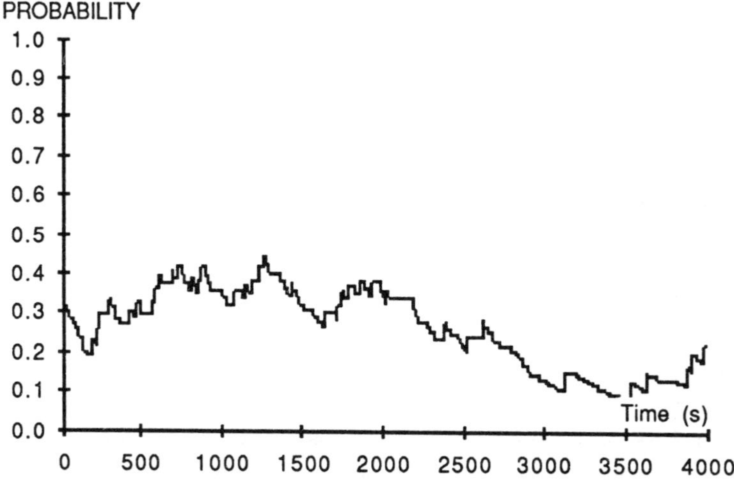

Figure 42. Probability evolution (first level, second interval)

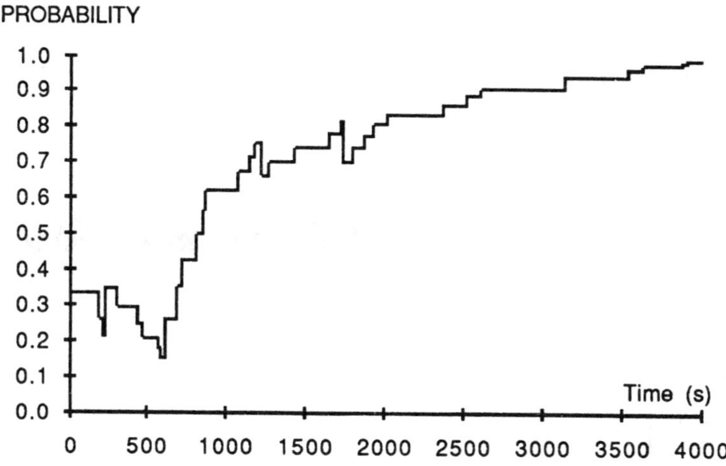

Figure 43. Probability evolution (second level, second automaton, third interval)

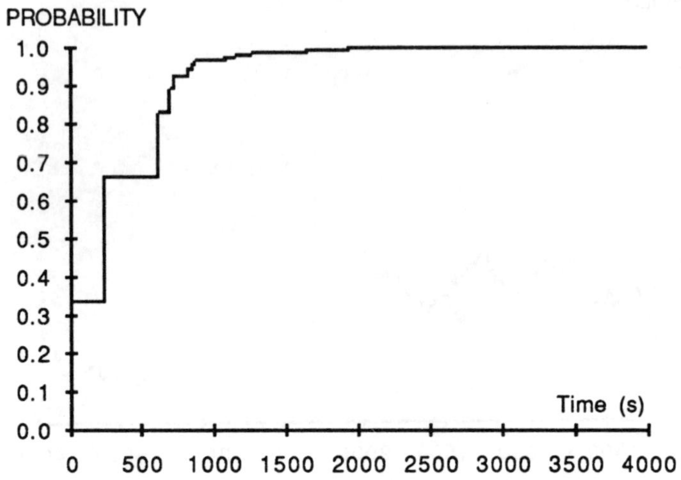

Figure 44. Probability evolution (third level, sixth automaton, third interval)

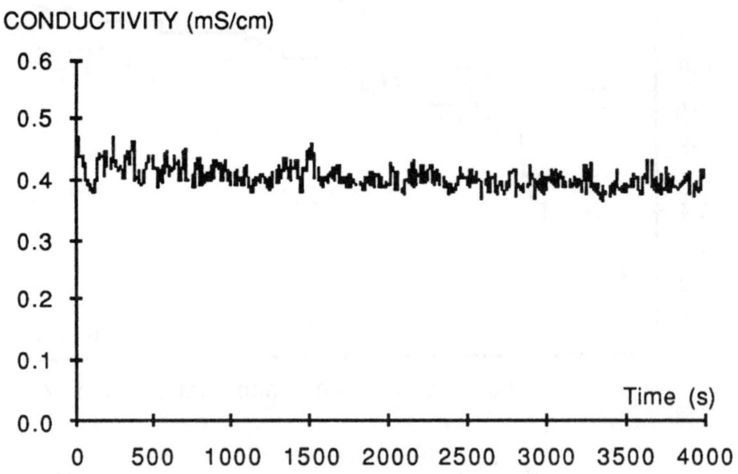

Figure 45. Time evolution of conductivity

The following experiment was carried out using an aqueous solution of 5 weight % of acetone as the continuous phase. The desired value of the conductivity was 0.40 mS/cm.

The same representations as for the previous experiment are given as figures 45, 46, 47 and 48 for the conductivity, the pulse frequency, the continuous phase and dispersed phase flow rate, respectively.

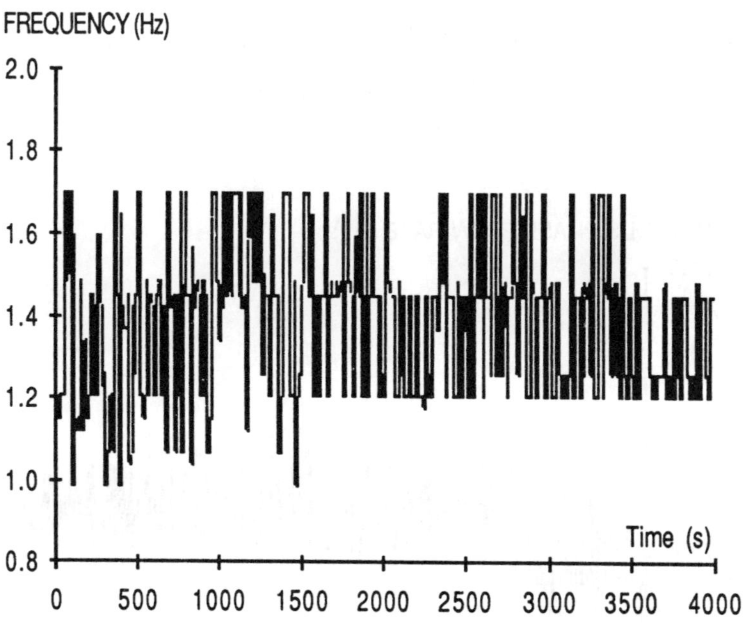

Figure 46. Time evolution of pulse frequency

A sampling period of 10 seconds was chosen with respect to the step response of the conductivity following a change in the pulse frequency (figure 2). The value of the conductivity corresponding to the very beginning of flooding is 0.45 mS/cm (desired value related to the optimal behaviour of the column.

It can be considered that the conductivity (not filtered) is maintained close to the desired value in spite of fluctuations. These fluctuations correspond to an abrupt change of the dispersed feed flow rate (the continuous phase flow rate also fluctuates) and strengthens the idea of a strong coupling between the variables characterizing the column behaviour. Nevertheless, the control objective have been reached succesfully.

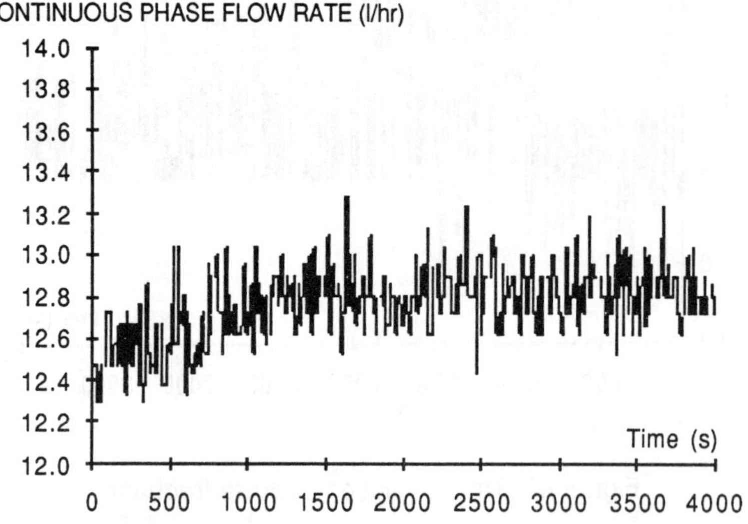

Figure 47. Time evolution of continuous phase flow rate (water+acetone)

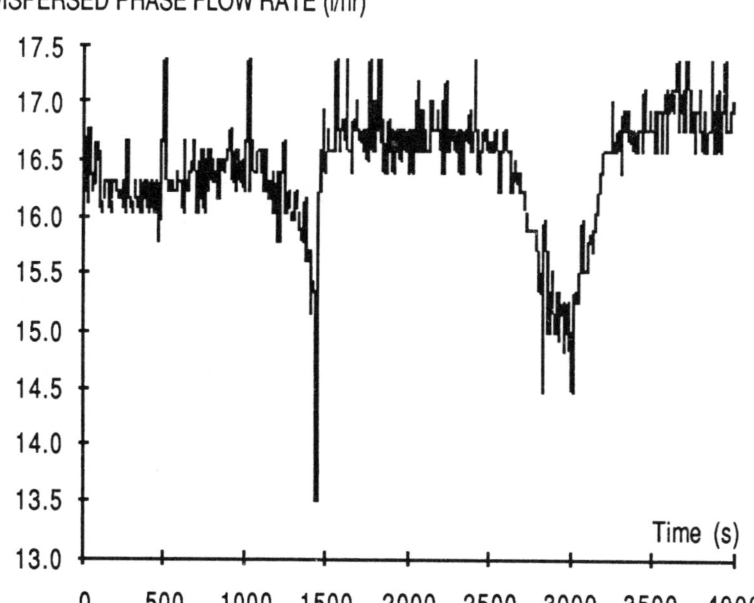

Figure 48. Time evolution of dispersed phase flow rate (toluene)

Figures 49, 50 and 51 give an example of probability evolution for each level:

- first level: probability related to the 2nd interval,
- second level: probability related to the 1st interval of the 2nd automaton,
- third level: probability related to the 2nd interval of 4th automaton, respectively.

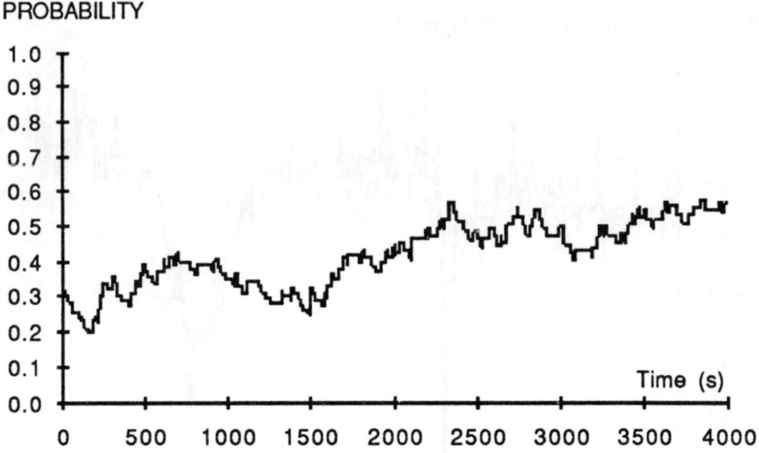

Figure 49. Probability evolution (first level, second interval)

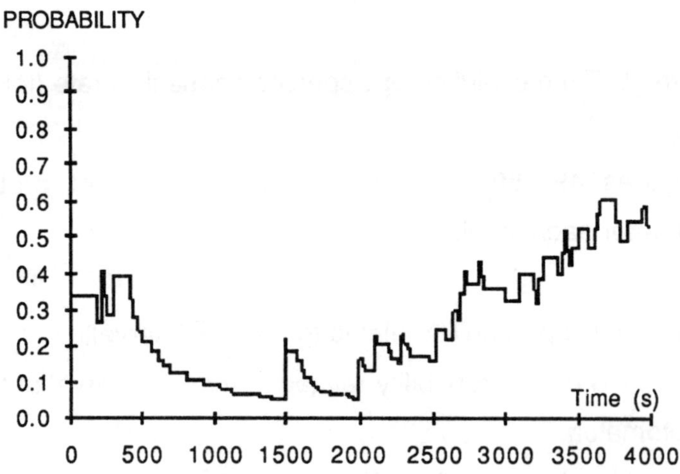

Figure 50. Probability evolution (second level, second automaton, first interval)

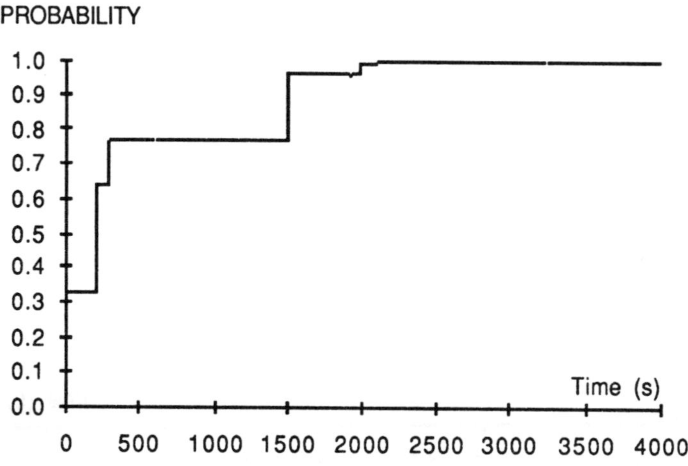

Figure 51. Probability evolution (third level, fourth automaton, second interval)

One can get information about the satisfactory behaviour of the column by looking at the variation of the concentration of acetone in the continuous phase outlet depicted in figure 52. After the period of time related to start-up and to the long transient time associated with the column response through the concentration of acetone in the continuous phase, the latter is close to the desired value.

Industrial processes are subject to varying conditions of operation and under the increasing need for flexibility and polyvalence, the interest of adaptive control becomes more and more evident. The principal requirements for such controllers are obvious: on one hand, their ability to adapt themselves to

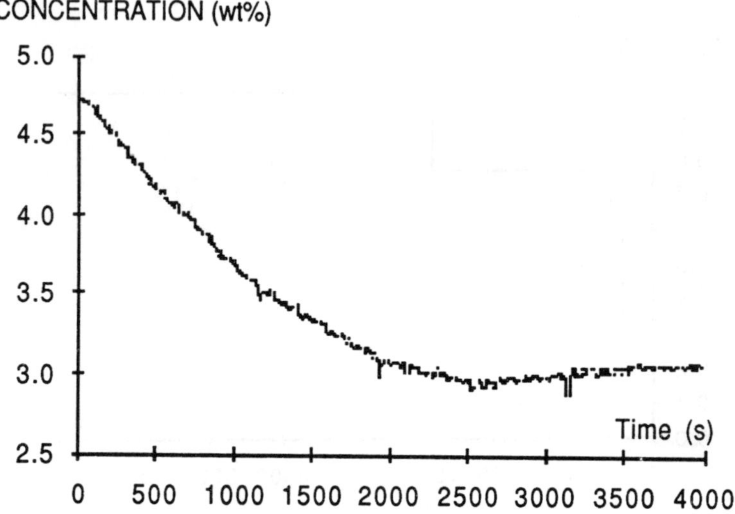

Figure 52. Time evolution of the concentration of acetone in the
continuous phase outlet

unexpected perturbations or varying operation conditions and on
the other hand the necessity of minimal a priori information about
the process.

As far as the chemical industry is concerned, and among the
various types of unit operation equipment, liquid-liquid extraction
columns exhibit complex, non-linear and representative
behaviours and they reveal themselves a valuable tool for
extensive studies

In this paragraph, it is shown that a simple control design
based on a multilevel system of automata gives very satisfactory
control results. This approach has a particular advantage in that it

requires no tuning at all.

The next section will be concerned with major improvements expected, including implementation of multivariable algorithms.

8. MULTIVARIABLE LEARNING CONTROL

8.1. SIMPLE APPROACH

A second loop of regulation has been introduced in order to maintain the solvent consumption at the minimum level necessary to achieve a given rate of mass transfer, i.e., to maintain a given solute concentration in the outlet raffinate phase, while the column is controlled at the boundary of flooding by means of the technique described above.

Therefore, the conductivity of the medium measured under the distributor (first loop) and the concentration of acetone in the raffinate (second loop) are both selected as the controlled variables. The pulse frequency and the dispersed solvent flow rate are chosen as the control actions (Najim et al. 1987c).

As in section 5 two pyramidal structures of automaton are implemented for control purpose. The schematic diagram of this control strategy is given in figure 53.

The know-how on the process introduced in the performance evaluation unit has been reduced to the minimum and can be summarized as follows:

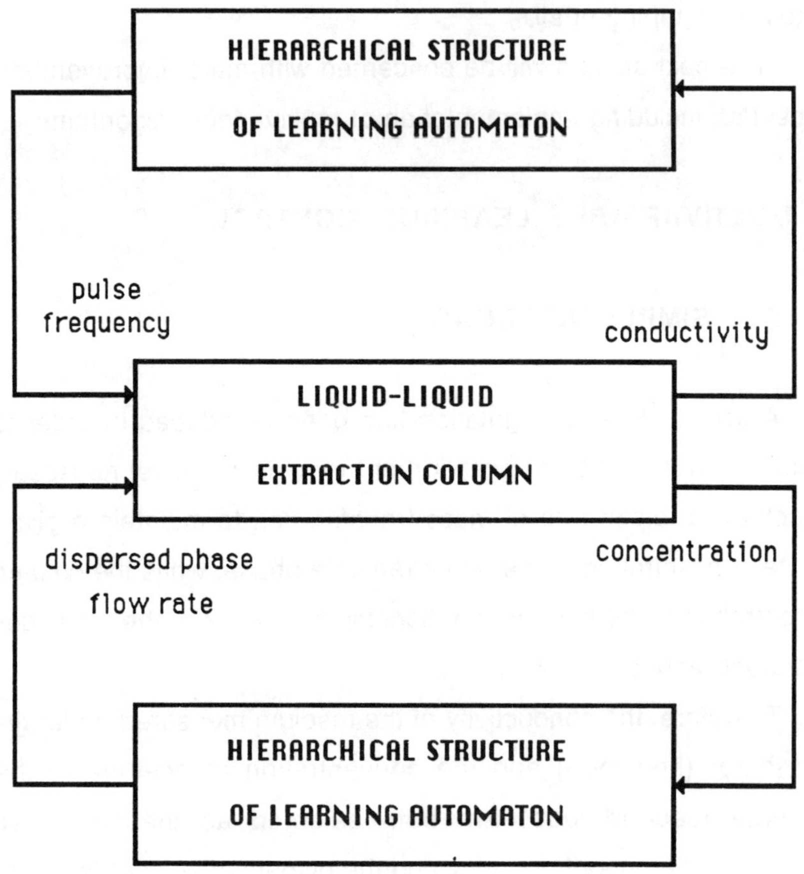

Figure 53. Control diagram using two hierarchical structure of automaton

first loop:

-if the conductivity measure is above the reference, the pulse frequency must be increased to reach flooding conditions.

-if the conductivity measure is below the reference, the pulsation must be decreased.

This leads to the mathematical formulation:

$$\underline{w(t) = 0} \quad \text{if } (y(t) < yc \text{ and } u(t) > u(t - 1)$$
$$\text{or } (y(t) > yc \text{ and } u(t) < u(t - 1)) \tag{5}$$
$$\underline{w(t) = 1} \quad \text{otherwise}$$

where yc is the desired conductivity.

second loop:

-in the same way, two rules describing the minimum a priori information on the column have been established:

-if the concentration of acetone in the raffinate is greater than the desired value, the solvent feed flow rate must be increased to extract more acetone from the aqueous solution.

-in the other case, the solvent feed flow rate must be decreased.

These rules can be mathematically formulated in a similar way to the previous one.

Figures 54 and 55 indicate respectively the time variation of the conductivity and the concentration in the continuous phase (water + remaining acetone) outlet. The conductivity increase at time t = 1700 seconds is consecutive to a decrease of both pulse frequency and solvent feed flow rate.

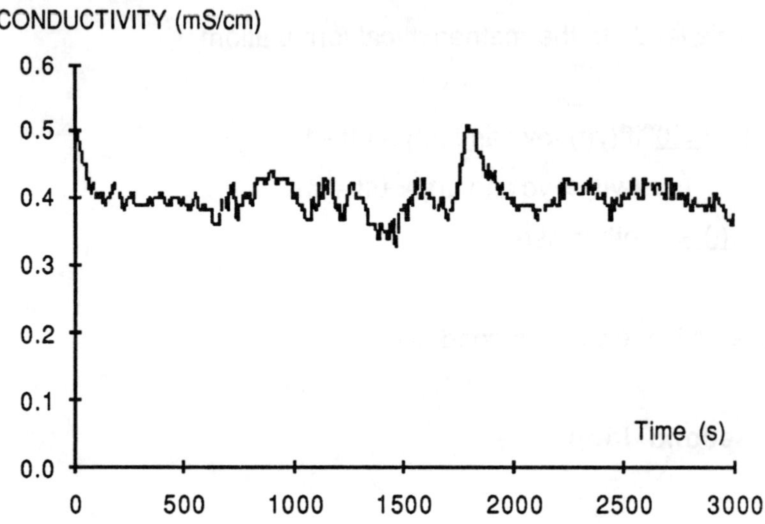

Figure 54. Time evolution of conductivity

The concentration of acetone in the raffinate is kept close to the desired value 3%.

Another approach consists of using two hierarchical structures of automaton where each automaton receives some information from the other. This information will be composed of the input and the output related to each single control loop as is indicated in figure 56. The connection between these automata can appear in the reward and penalty formulation.

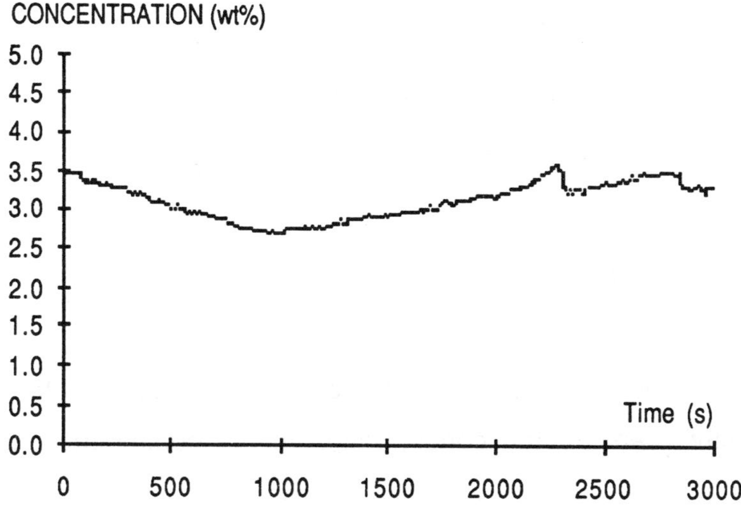

Figure 55. Time evolution of the concentration of acetone in the continuous phase outlet

$\underline{w(t) = 0}$ (first loop)

if $\{(y_2(t) > yc_2)$ and $(u_2(t) > u_2(t-1))$ and $(y_1(t) > yc_1)\}$ or

$\{(y_2(t) > yc_2)$ and $(u_2(t) < u_2(t-1))$ and $(y_1(t) < yc_1)\}$ or $\qquad$ (6)

$\{(y_2(t) < yc_2)$ and $(u_2(t) < u_2(t-1))$

or $((y_2(t) = yc_2)\}$ and $(u_2(t) = u_2(t-1))$

$\underline{w(t) = 0}$ (second loop)

if $\{(y_1(t) > yc_1)$ and $(u_1(t) > u_1(t-1))\}$ or

$$\{(y_1(t) < yc_1) \text{ and } (u_1(t) < u_1(t-1))\} \text{ or}$$

$$\{(y_1(t) = yc_1) \text{ and } (u_1(t) = u_1(t-1))\} \text{ or} \tag{7}$$

$$\{(y_1(t) > yc_1) \text{ and } (u_1(t) < u_1(t-1)) \text{ and } (u_2(t) > u_2(t-1))\}$$

$\underline{w(t) = 0}$ otherwise

The indexes 1 and 2 correspond respectively to the conductivity and to the concentration in the continuous phase outlet.

Figures 57 and 58 show the time variation of, respectively, the conductivity and the concentration in the continuous phase (water + remaining acetone) outlet.

These experiments demonstrate that the presented algorithm has excellent control properties. The conductivity and the concentration of acetone in the continuous phase are close to the desired values with high precision (equal to the precision of the sensors used).

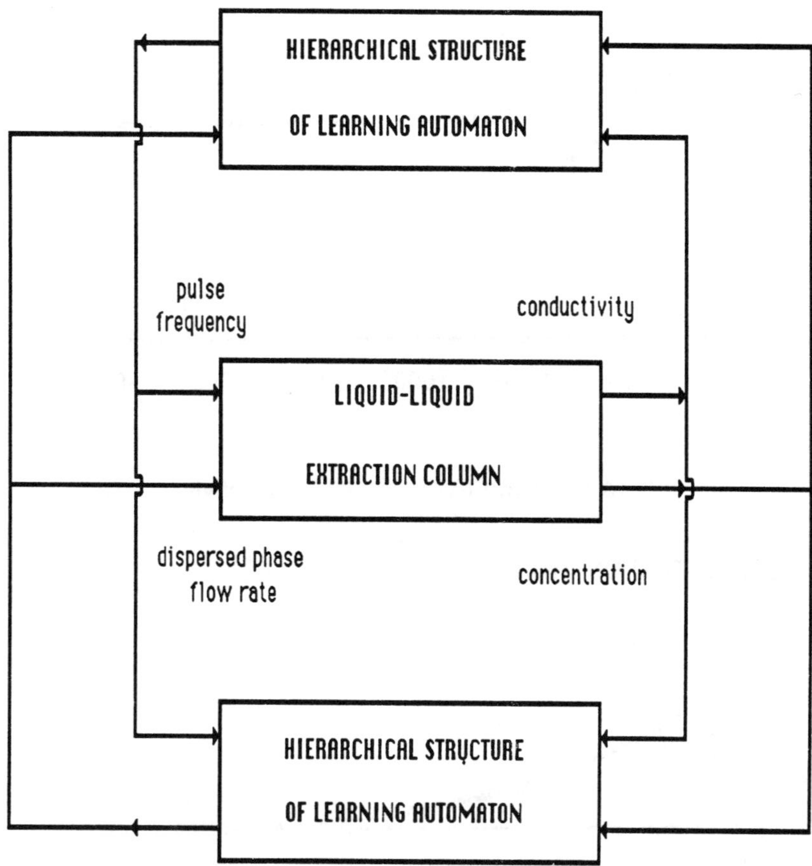

Figure 56. Control diagram using two interconnected
hierarchical structures of automaton

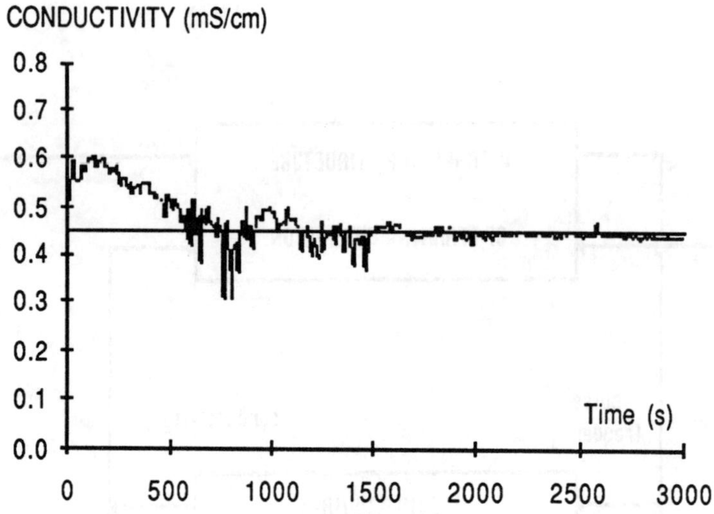

Figure 57. Time evolution of conductivity

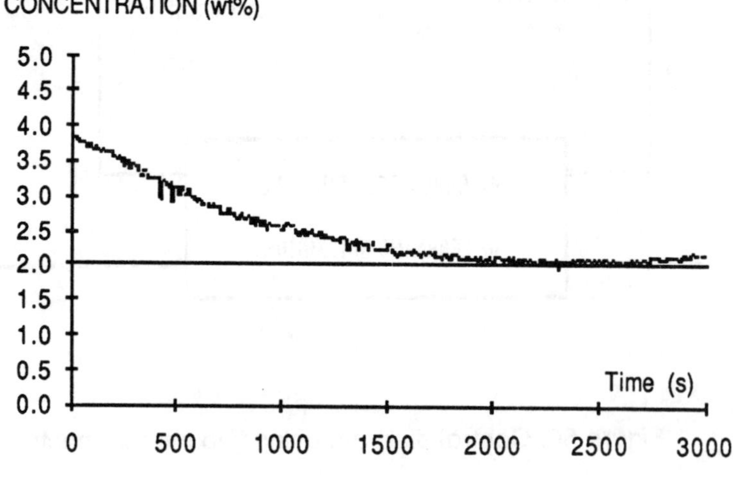

Figure 58. Time evolution of the concentration of acetone in the continuous phase outlet

8.2. GLOBAL APPROACH

A single hierarchical structure of automata (N levels) was used to control the conductivity and the concentration of acetone in the raffinate (figure 59).

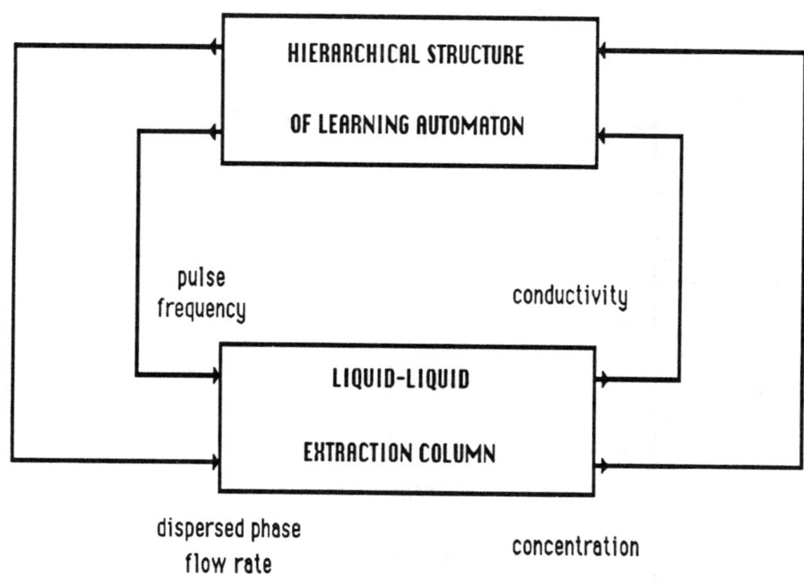

Figure 59. Multivariable control diagram using one hierarchical structure of automaton

The pulse frequency and the dispersed flow rate domains variation are respectively discretized into a set of M values $(N^N = M^2)$:

The pulse frequency F and the dispersed phase flow rate Q_d take values in the domains $[f_1, f_2,....., f_M]$ and $[q_1, q_2,......, q_M]$ where $f_1 = f_{min}$, $f_M = f_{max}$, $q_1 = Q_{dmin}$ and $q_M = Q_{dmax}$. With $f_{min} = 0.96$ Hz, $f_{max} = 1.73$ Hz, $Q_{dmin} = 8$ l/h, $Q_{dmax} = 12$ l/h.

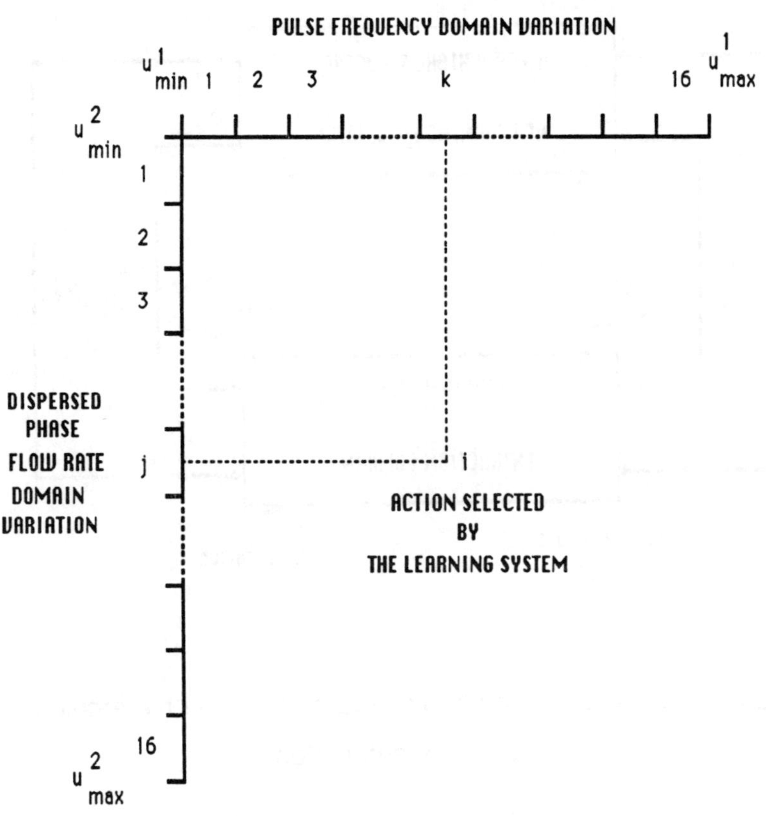

Figure 60. Correspondence between the control action u_i and the pulse requency and the dispersed flow rate.

The control action u_i selected at time t corresponds to a couple:

Pulse frequency = f_j ; $j = i - (k - 1)*M$ $(N = 4 ; M = 16)$

Dispersed flow rate = q_k ; $k = 1 + (i - 1)/M$

This correspondence is explained by the diagram given in figure 60.

The performance evaluation unit generates reward or inaction on the basis of simple combination of the rules (relations 6 and 7).

The conductivity and the concentration in the continuous phase (water+ remaining acetone) outlet are depicted in figures 61 and 62 repectively.

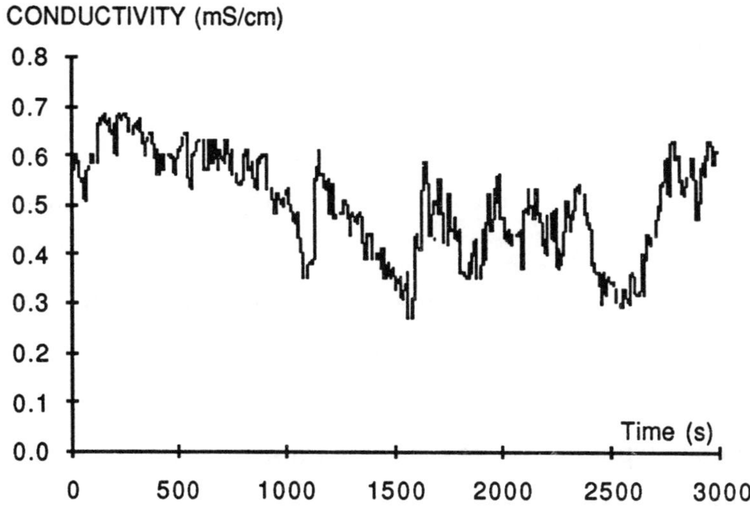

Figure 61. Time evolution of conductivity

The conductivity is more disturbed than in the previous case. Its mean value is close to the desired one. The evolution of the concentration is excellent. After approximately 800 seconds, the concentration reaches the desired value and becomes stable. From an industrial view point, this output is the most important variable to be considered.

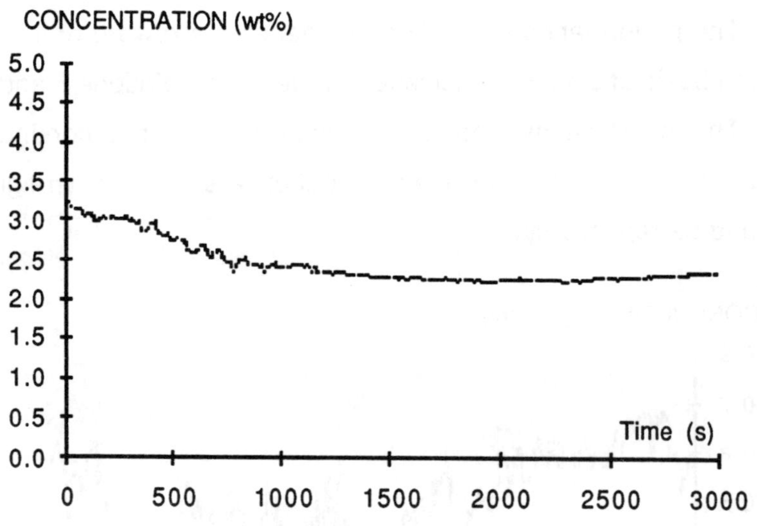

Figure 62. Time evolution of the concentration of acetone in the continuous phase outlet

These experimental results illustrate the efficiency of this control approach.

The learning control system has been developed with a very poor a priori knowledge of the plant introduced in the performance evaluation unit by means of very simple rules. More complex heuristic rules could be used and consequently could improve the control quality. It is noticed that in all the previous experiments the measurements were not low-pass filtered. The control performances can be also improved by selecting two different sampling periods, i.e., one for the conductivity which will noted T_1 and one for the concentration which will be noted T_2. T_1 may be great than T_2 ($T_2 = \alpha\, T_1$; α is an integer number) . Najim and Al Khani (1988) have obtained good results in the application of a multi-input multi-output generalized minimum variance controller with the following choice of sampling periods: $T_1 = 10$ s, $T_2 = 40$ s.

9. KÜHNI COLUMN DESCRIPTION

The Kühni column is a major representative of rotary mixed columns. The experimental pilot plant is made of glass of 125 cm in height and 10 cm in diameter. Internals in the column are sieve plates made of stainless steel and turbines; the dimensions of internals are according to Khüni's standard normalization. The stirring turbines are fixed on a central rotating shaft connected with an electric engine at the top of the column. The perforated plates are fixed : each turbine is exactly centered between two plates, an arrangement of two plates and one turbine determining one mixing compartment. The Kühni diagram is shown in figure 63.

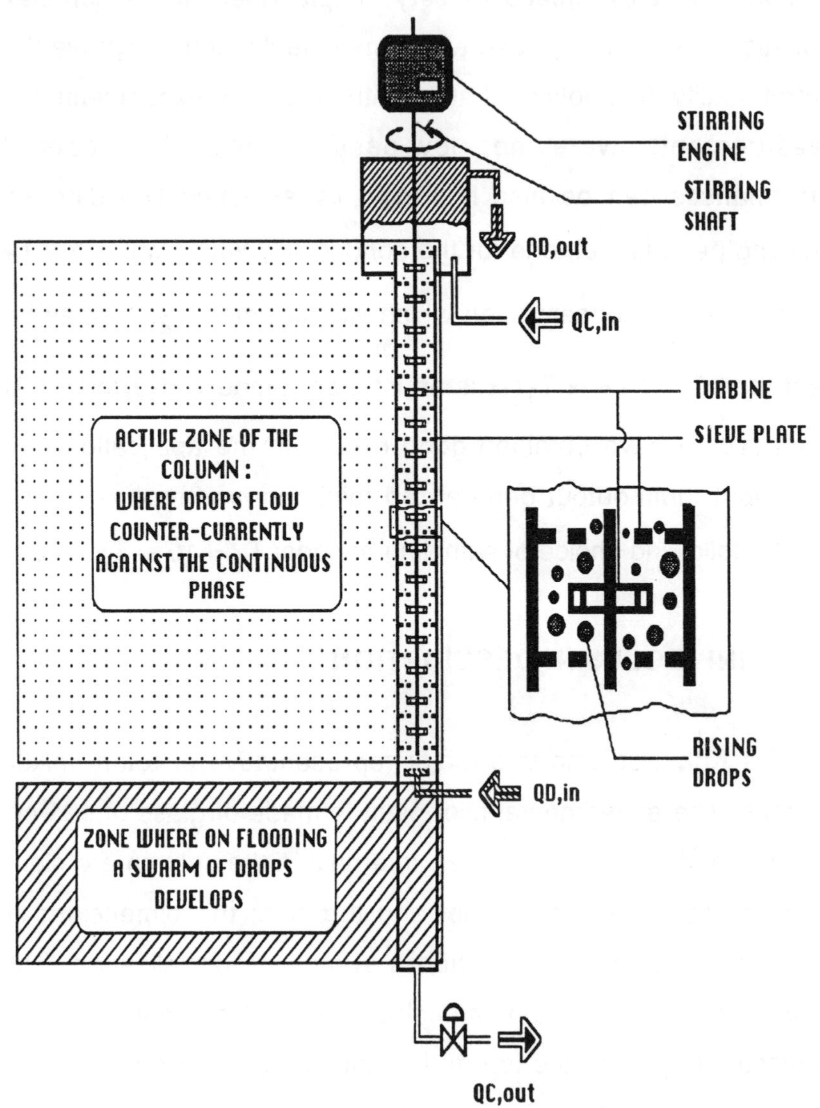

Figure 63. Schematic diagram of a Kühni column

As in the pulsed column, the ternary system used is water-acetone-toluene. The solvent phase is the light phase and is dispersed at the bottom of the column by means of a perforated pipe distributor. Drops of the dispersed phase rise through the descending continuous water phase and decant at the top of the column in a settling zone. The controlled variable remains the conductivity of the liquid medium measured at a place located between the dispersed phase inlet and the continuous phase outlet, while the pulsing frequency has been replaced by the rotation speed of the central shaft which carries the mixing turbines.

The dynamics of the Kühni column is as complex as those related to the pulse liquid-liquid extraction columns, and there exists a great analogy between their behaviour.

Only the experimental results obtained from the the application of generalized predictive control algorithm with multiple model reference to the Kühni column will be presented in the following paragraph. The learning control design for this kind of column is similar to the pulsed column, and lead to similar results.

As for pulsed columns, apart from the properties of liquid-liquid systems, four main factors affecting the Kühni column performance may be considered: rotation speed of the central shaft, feed flow rates, phase flow ratio, and concentration of the aqueous and solvent phases.

10. PREDICTIVE CONTROL OF A KÜHNI COLUMN

Figure 64 shows the conductivity variation following a change in the rotation speed of the central shaft (from 125 to 242 rpm).

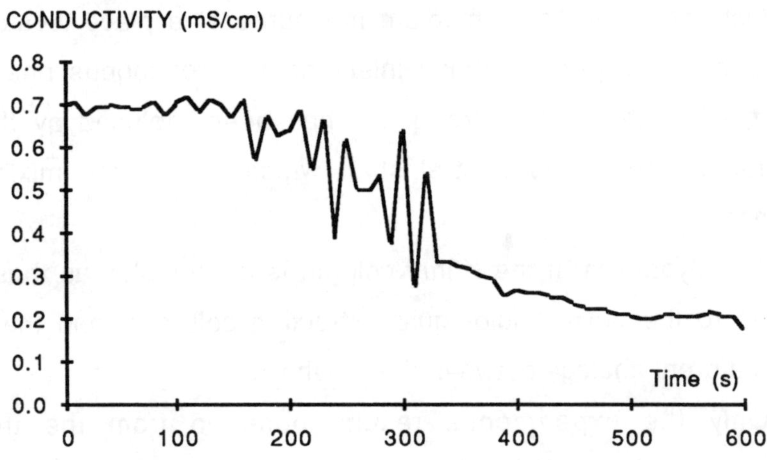

Figure 64. Step response of the Kühni column

From this step response a suitable model order, time delay and an appropriate sampling period were determined. The model structure obtained is similar to the dynamics representation of the pulsed column.

The control objective was to maintain the conductivity around the desired value: about 0.60 mS/cm, which corresponds to the very beginning of flooding.

The modified generalized predictive control algorithm presented in the previous chapter has been implemented on an Apple II to control the considered Kühni column (Najim et al., 1987a).

The process parameters are unknown and time-varying, they are recursively estimated and the control strategy is developed in self-tuning manner by replacing the process parameters by their estimates. The model's parameters are initially identified by decreasing gain with forgetting factor. Once the trace of the matrix is below a certain level then the constant trace algorithm is used. The initial values associated with this algorithm are the same as those used for the pulsed column control.

The characteristic polynomial $P(q^{-1})$ was chosen to be equal: $P(q^{-1}) = (1 - 0.45q^{-1})^2$.

The following experiments were carried out to maintain the column in its optimal behaviour zone, in spite of disturbed flow rates.

Good results were obtained, as shown in figures 65, 66, 67 and 68 which represent the variations of the conductivity, the speed rotation, the continuous phase and the dispersed phase flow rates respectively.

In spite of little prior knowledge introduced in the control algorithm, the conductivity is close to the desired value (0.6 m S/cm) after a short transient time (about 40 seconds). The regulation error (difference between the measured output and the set point) is less than 1.5%.

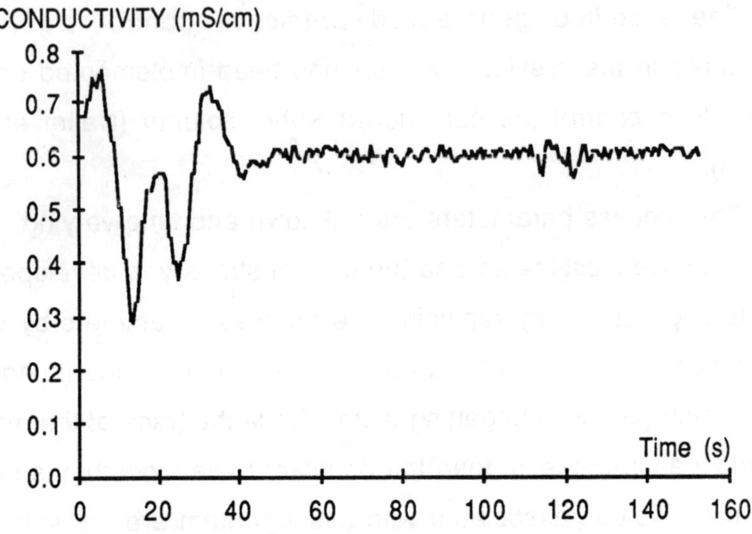

Figure 65. Time evolution of conductivity

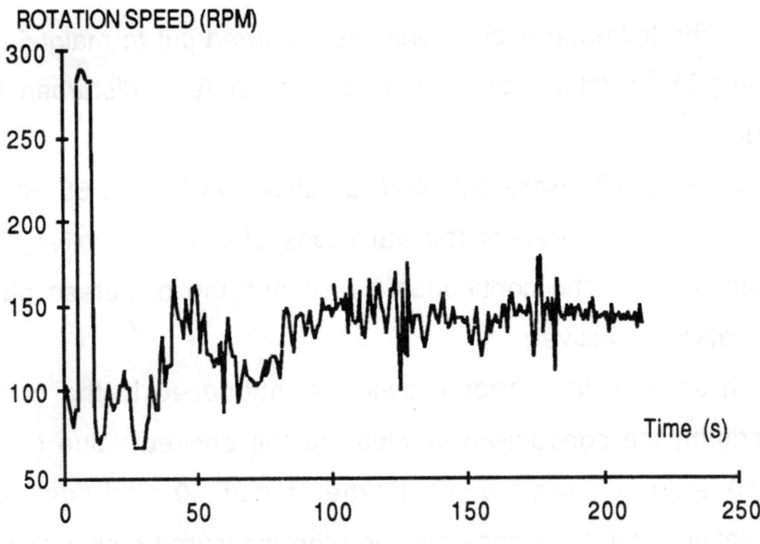

Figure 66. Time evolution of speed rotation

This measured output was not filtered. The control action has relatively smooth variations. The fluctuations are due to noise affecting the conductivity measurements which can be easily avoided by using a simple low pass filter.

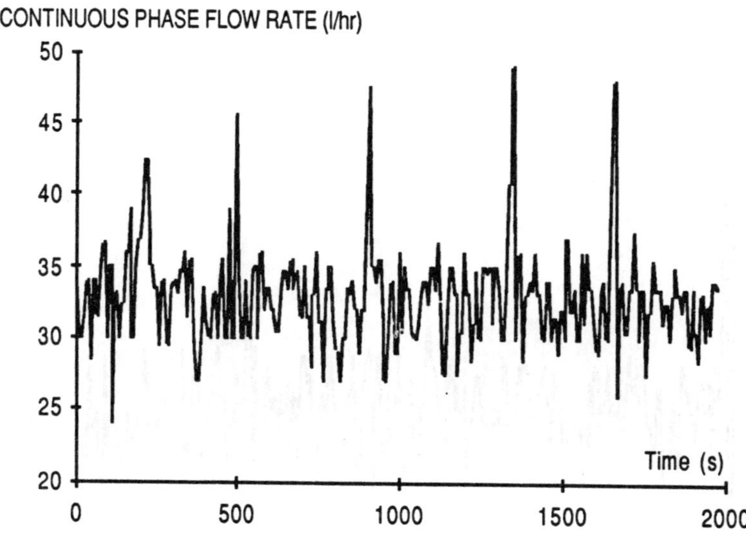

Figure 67. Time evolution of continuous phase flow rate (water)

In spite of the high fluctuations of the different flow rates, the adaptive algorithm implemented brings the considered column to its optimal behaviour zone and maintains it there. The results obtained can be improved by conditioning the conductivity measurements.

Another experiment was carried out with an aqueous solution of 6.65 (vol%) of acetone as continuous phase, with the same conductivity reference (0.60 mS/cm). Figures 69, 70,71 and 724 give the time evolution of the output variable (conductivity), the control variable (rotation speed) and both flow rates (continuous and dispersed phase) respectively.

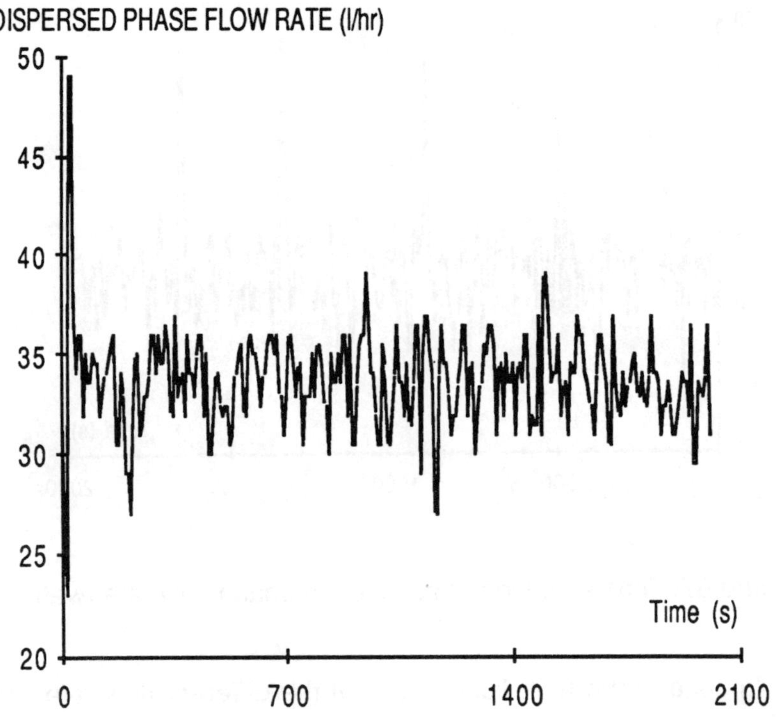

Figure 68. Time evolution of dispersed phase flow rate (toluene)

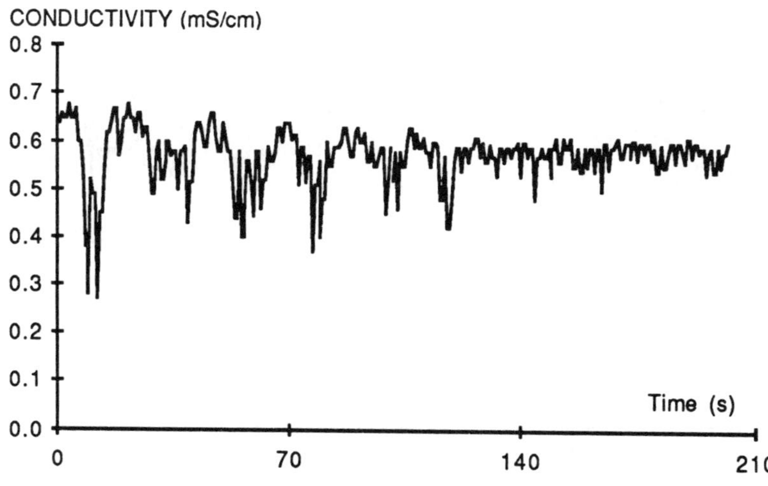

Figure 69. Time evolution of conductivity

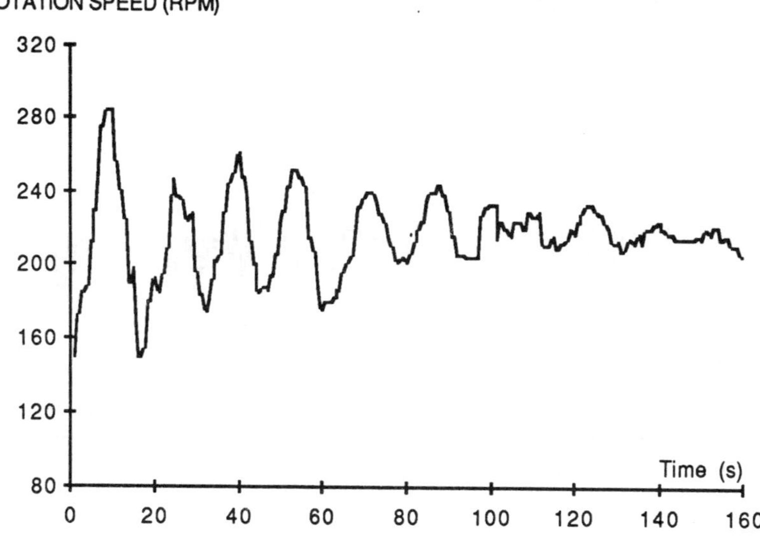

Figure 70. Time evolution of speed rotation

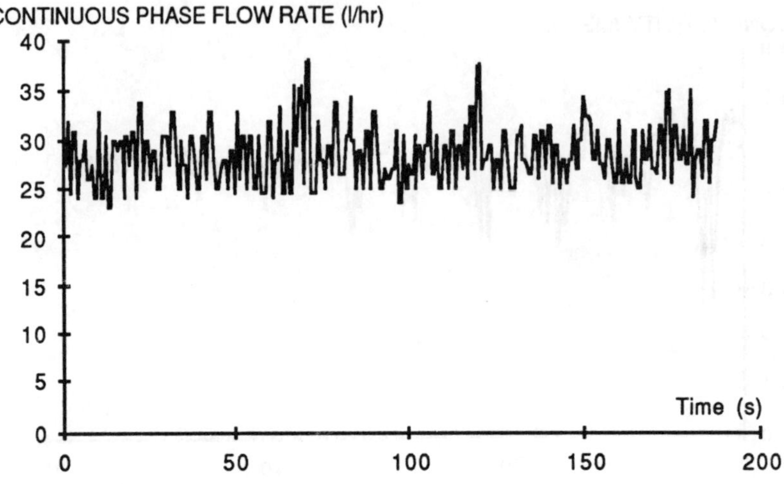

Figure 71. Time evolution of continuous phase flow rate (water)

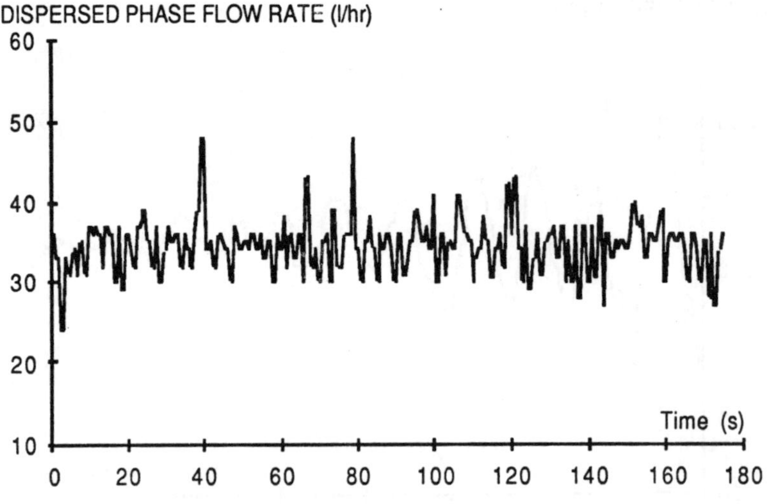

Figure 72. Time evolution of dispersed phase flow rate (toluene)

An indication of the good behaviour of the column is given by the amount of acetone remaining in the continuous phase outlet, as is shown in figure 73.

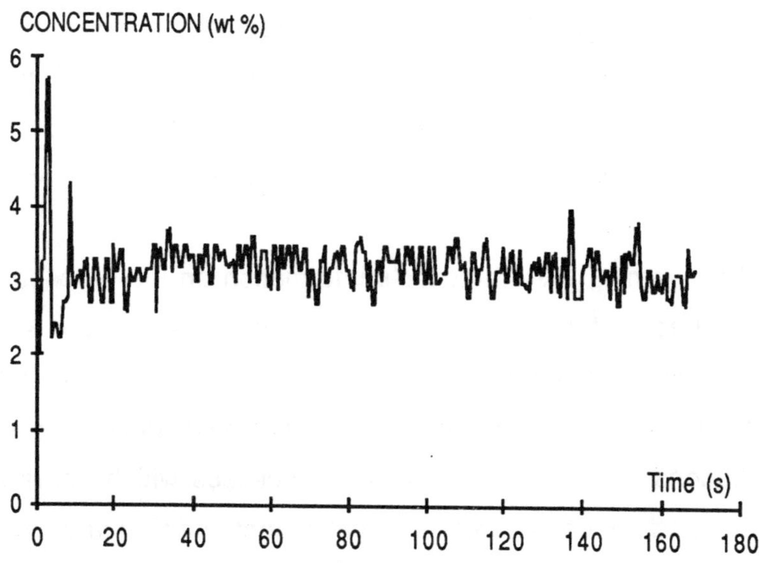

Figure 73. Time evolution of the concentration of acetone in the continuous phase outlet

11. CONCLUSION

The regularized pole placement, the linear quadratic Gaussian controller control techniques, the associated of long-range predictive control algorithm with the model reference techniques, and the learning methods are very attractive, robust and easy to implement for control purposes on a digital computer which has a limited computing capacity like a low-cost microprocessor. On applying these algorithms to various mechanically agitated colums, valuable results have been obtained. These control methods do not require the elaboration of an a priori model of the columns which generally exhibit highly non-linear terms and are adapted to real-time control. The long-range predictive control algorithms are relatively insensitive to parametrization.

The learning system collects information on the process and on the basis of heuristic rules chooses values of the action variables to bring the considered process to its desired operating point and keep it there. This control approach may have wide applications in the field of industrial processes which, more and more, need to be flexible and polyvalent, particularly in fine chemistry.

The results presented show the ability of the adaptive control algorithms to deal with the control of complex systems with time varying dynamics. The control approaches are general and can be used to control any chemical or biotechnological process. They may be used even by those who are unfamiliar with the myriad details that make the presented tools efficient.

APPENDIX A

LINEAR QUADRATIC GAUSSIAN CONTROLLER

For the process described by the following state equation

$$x(n + 1) = A\, x(n) + B\, u(n) + w(n)$$

$$y(n) = C\, x(n) + v(n)$$

(A.1)

The stationary white noise sequences $w(n)$ and $v(n)$ are assumed to be of zero mean and independent, and to have the covariance matrices Q_n (semidefinite positive matrix[*]) and R_n (positive definite matrix[*]).

Based on the separation principle, the synthesis of the the linear quadratic Gaussian controller will be derived in two steps. The first step concerns the resolution of the deterministic linear quadratic control problem, and the second step is related to state estimation using the Kalman filter.

[*]   A matrix R is definite positive if <a, Ra> is positif for all a ≠ 0 and a matrix Q is positive semidefinite if and only if <a, Qa> ≥ 0. The scalar product <a, b> is given by $\Sigma\ a_i b_i$ where b_i and a_i are respectively the components of the vectors a and b.

Let us consider the deterministic linear quadratic control problem which can be stated as follows: derive the control law for calculating the control actions which minimize the following cost function:

$$J = \{\sum_{n=1}^{N} [x^T(n)Qx(n) + u^T(n)Ru(n)]\}/2 \tag{A.2}$$

Q is a positive semidefinite matrix (symmetric)

R is a positive definite matrix (symmetric)

This cost function generally defines the objective of a control system from a physical point of view.

The free noise model (equation (A.1)) may be considered as a constraint for this optimization problem. This constraint (- x(n + 1) + A x(n) + B u(n) = 0) will be handled by using the Lagrange multipliers method (Athans and Falb, 1966).

The control criterion may be written as:

$$J_L = \{\sum_{n=1}^{N} [x^T(n)Qx(n) + u^T(n)Ru(n)] + \lambda^T(n + 1)[- x(t + 1) + A x(t) + B u(t)] \}/2 \tag{A.3}$$

The cost function J_L will be minimized toward x(n), u(n) and λ(n).

Classic minimization of J_L leads to the solution of this optimal control problem. On the optimal trajectory, the following equations must be verified:

control equation

$$\partial J_L / \partial u(n) = u^T(n)R + \lambda^T(n+1)B = 0 \qquad (A.4)$$

state equation

$$\partial J_L / \partial \lambda(n) = -x(tn+1) + A\,x(n) + B\,u(n) = 0 \qquad (A.5)$$

adjoint equation

$$\partial J_L / \partial x(n) = x^T(n)Q - \lambda^T(n) + \lambda^T(n+1)\,A = 0 \qquad (A.6)$$

Equation (A.6) gives the Lagrange multipliers.

$$\lambda(n) = A^T \lambda(n+1) + Q\,x(n) \qquad (A.7)$$

The optimal solution in the sense of the minimization of the criterion defined above is given by the solution of the set of equations (A.4), (A.5) and (A.7).

Usually the initial state $x(0)$ is known; on the other hand λ_0 is unknown. To solve the optimization problem in which we are interested, let us consider the terminal conditions. If the minimum value of the quadratic cost function is reached, the control action $u(n)$ may be equal to zero. Taking into consideration the state space system representation (equation (A.1)), it follows that the control $u(n)$ does not bear on the state vector $x(n)$ (e.g., $x(n) = A$

x(n - 1) + B u(n - 1)). Then, from equation (A.4) we can set the terminal value (for n = N + 1) of the Lagrange multiplier to be equal to zero, i.e., $\lambda(N + 1) = 0$.

Equation (A.7) can be rewritten as

$$\lambda(N) = Q\, x(N) \tag{A.8}$$

Consequently, the solution of equations (A.5) and (A.7) constitutes the solution of the optimization control problem stated above. This problem, which is called the two-point boundary-value problem, is not easy to solve.

Let us attempt to find a solution of the form (assume that the Lagrange multiplier is related to the state variable x(n) by a linear function):

$$\lambda(n) = P(n)\, x(n) \tag{A.9}$$

Introducing (A.9) into (A.4), and using (A.1) we get

$$R\, u(n) = -\, B^T P(n + 1)\, [A\, x(n) + B\, u(n)\,] \tag{A.10}$$

Equation (A.10) now gives

$$u(n) = -\, \{[R + B^T P(n + 1)\, B\,]^{-1}\, B^T P(n + 1)\, A\,\}x(n) \tag{A.11}$$

Equation (A.11) shows that the control is a linear combination of the state vector components.

In order to present the result in a simple form, let R_1 denote the matrix sum defined by: $R_1 = [R + B^T P(n + 1)B]$. Then, equation (A.11) may be written as

$$u(n) = - [R_1]^{-1} B^T P(n + 1) A\, x(n) \tag{A.12}$$

Simple manipulations of the previous equations lead to

$$[\, P(n) - A^T P(n + 1)\, A + A\, P(n + 1)B(R_1)^{-1}B^T P(n + 1)A - Q\,]\, x(n) = 0$$

$$\tag{A.13}$$

This equation holds for any $x(n)$, then the recursion for calculating $P(n)$ is given by:

$$P(n) = A^T [P(n + 1) + P(n + 1)B(R_1)^{-1}B^T P(n + 1)]\, A + Q$$

To solve the linear quadratic control problem (deterministic or stochastic), other methods can be used. The calculus of variations and dynamic programming can be invoked respectively in the continuous and discrete cases.

APPENDIX B

RECURSION OF THE DIOPHANTINE EQUATIONS

We shall in this appendix present recursive methods related to the solution of the identities appearing in the predictive control algorithms.

1. THE IDENTITY: $1 = E_j(q^{-1})A(q^{-1})\Delta(q^{-1}) + q^{-j}F_j(q^{-1})$

Suppose that the polynomials $E_j(q^{-1})$ and $F_j(q^{-1})$ are known:

$$1 = E_j(q^{-1}) A(q^{-1}) \Delta(q^{-1}) + q^{-j} F_j(q^{-1}) \qquad (B1.1)$$

Consider the Diophantine Equation for j+1 then:

$$1 = E_{j+1}(q^{-1})A(q^{-1})\Delta(q^{-1}) + q^{-j-1} F_{j+1}(q^{-1}) \qquad (B1.2)$$

Substracting Eq. (B1.1) from Eq. (B1.2) gives:

$$0 = A \Delta (E_{j+1} - E_j) + q^{-j} (q^{-1} F_{j+1} - F_j) \qquad (B1.3)$$

The polynomial $E_j (q^{-1})$ being of degree j-1, then the difference

$[E_{j+1}(q^{-1}) - E_j(q^{-1})]$ is of order j. Therefore, it may be written as:

$$E_{j+1}(q^{-1}) - E_j(q^{-1}) = R(q^{-1}) + r_j q^{-j} \tag{B1.4}$$

So that:

$$0 = A(q^{-1})\Delta(q^{-1})R(q^{-1})+$$
$$+ q^{-j}\{q^{-1}F_{j+1}(q^{-1}) - F_j(q^{-1}) + A(q^{-1})\Delta(q^{-1})r_j\} \tag{B1.5}$$

By examining the degree of polynomials appearing in equation (B.5) one may conclude that the solution of this equation is:

$$R(q^{-1}) = 0$$

and $\tag{B1.6}$

$$F_{j+1}(q^{-1}) = q[F_j(q^{-1}) - A(q^{-1})\Delta(q^{-1})r_j]$$

Term to term identification of q^{-i} coefficients gives:

$$r_j = f_{j,0}$$
$$f_{j+1,i} = f_{j,i+1} - a^*_{i+1} r_j \qquad\qquad i = 0,....,na \tag{B1.7}$$

where $f_{j,i}$, a^*_i are respectively the ith coefficients of the polynomial $F_j(q^{-1})$ and $\{A(q^{-1})\Delta(q^{-1})\}$.

The initial condition will be derived as follows.

For $j = 1$, then we have:

$$1 = E_1(q^{-1})A(q^{-1})\Delta(q^{-1}) + q^{-1}F_1(q^{-1}) \tag{B1.8}$$

To verify this identity, the first coefficient of the polynomial $E_1(q^{-1})$ must be equal to one ($a^*_1 = 1$), and the degree of q^{-1} $F_1(q^{-1})$ must be equal to the degree of the polynomial

$\{A(q^{-1})\Delta(q^{-1})\}$. Consequently, the degree of the polynomial $E_1(q^{-1})$ is equal to zero.

It follows that:

$$E_1(q^{-1}) = 1$$

$$F_1(q^{-1}) = q\{1 - A(q^{-1})\ \Delta(q^{-1})\}$$

2. THE IDENTITY: $E_j(q^{-1})B(q^{-1}) = G^{\circ}_j(q^{-1})F_f(q^{-1}) + q^{-j}G^{*}_j(q^{-1})$

Let us consider the identity mentioned above for j and j+1

$$E_j(q^{-1})B(q^{-1}) = G^{\circ}_j(q^{-1})F_f(q^{-1})\ q^{-j}G^{*}_j(q^{-1}) \qquad (B2.1)$$

$$E_{j+1}(q^{-1})B(q^{-1}) = G^{\circ}_{j+1}(q^{-1})F_f(q^{-1})\ q^{-j-1}G^{*}_{j+1}(q^{-1}) \qquad (B2.2)$$

With

$$E_j(q^{-1}) = e_0 + e_1 q^{-1} + ... + e_{j-1}q^{-j+1} \qquad (B2.3)$$

$$G^{\circ}_j(q^{-1}) = g^{\circ}_0 + g^{\circ}_1 q^{-1} + ... g^{\circ}_j q^{-j} \qquad (B2.4)$$

Substracting (B2.1) from (B2.2) leads to:

$$\{E_{j+1}(q^{-1}) - E_j(q^{-1})\}B(q^{-1}) =$$

$$\{G^{\circ}_{j+1}(q^{-1}) - G^{\circ}_j(q^{-1})\}F_f(q^{-1}) + q^{-j}\{q^{-1}G^{*}_{j+1}(q^{-1}) - G^{*}_j(q^{-1})\} \qquad (B2.5)$$

From equations (B2.3) and (B2.4), equation (B2.5) may be written as:

$$e_j q^{-j}B(q^{-1}) = g^{\circ}_{j+1}q^{-j}F_f(q^{-1}) + q^{-j}\{\ q^{-1}G^{*}_{j+1}(q^{-1}) - G^{*}_j(q^{-1})\} \qquad (B2.6)$$

By identifying term by term the coefficient of q^{-i}, we obtain the following relations:

$$g^{\circ}_{j+1} = e_j b_0 + g^{*}_{j,0} \qquad (B2.7)$$

$$g^{*}_{j+1,i-1} = g^{*}_{j,i} + e_j b_i - g^{\circ}_{j+1} f_i \ ;$$

$$i = 1, \max[\text{degree}B(q^{-1}), \text{degree}F_f(q^{-1})] \qquad (B2.8)$$

where, b_i and f_i are the coefficients of the polynomial $B(q^{-1})$ and $F_f(q^{-1})$.

The parameter $g^{*}_{j,i}$ represents the ith coefficient of the polynomial $G^{*}_{j}(q^{-1})$.

REFERENCES

Anderson B.D.O. and Moore J.B. (1979) "Optimal filtering", Prentice-Hall, Inc., New Jersy.

Anderson P. (1983) "Adaptive forgetting in recursive identification through multiple models", Department of electrical engineering report, Linköping university, code: LiTH-ISY-I-0638.

Åström K.J. (1970) "Introduction to stochastic control theory", Academic-Press, New York.

Åström K.J., and Wittenmark B. (1973) "On self-tuning regulator", Automatica, vol. 9, p.185.

Åström K.J. (1983) "Theory and applications of adaptive control-A survey", Automatica, vol. 19, n° 5, p. 471.

Åström K.J., and Wittenmark B. (1984) "Computer controlled systems theory and design", Prentice-Hall.

Åström K.J. (1987) "Adaptive feedback control", Proceedings of the IEEE, vol. 75, n° 2, p. 185.

Athans M. and Falb P.L. (1966) " Optimal control : An introduction to the theory and its applications", McGraw-Hill, New York.

Belanger P.R., Rochon L., Dumont G.A. and Gendron S. (1986) "Self-tuning control of chip level in a Kamyr digester", AIChE Journal, vol. 32, n° 1, p.65.

Bierman G.J. (1977) "Factorization methods for discrete sequential estimation", Academic Press, New York.

Clarke D.W., and Gawthrop J.J (1979) "Self-tuning control", Proc. IEE vol 126, n° 6, June, p. 633.

Clarke D.W. (1982) "Model following and pole-placement self-tuners", Optimal Applications & Methods, vol. 3, p.323.

Clarke D.W., Mohtadi C. and Tuffs P.S. (1984) "Generalized predictive control",
Part 1: The basic algorithm.
Part 2:Extensions and interpretations.
OUEL Reports Numbers 1555/84 - 1557/84.

Dahlqvist S.A. (1981)"Control of a distillation column using self-tuning regulators", Can. J. Chem. Engr., vol. 59, p.118.

De Keyser R.M.C. (1986) "Adaptive dead-time estimation", 2nd IFAC Workshop on Adaptive Systems in Control and Signal Processing, Lund, Sweden.

Desoer C.A. and Vidyasagar M. (1975) "Feedback systems: Input-ouput properties", Academic Press, New York.

Egardt B. (1979) "Stability of adaptive controllers", lecture notes in control and information sciences, n° 20, Berlin, Springer Verlag.

El-Fattah Y.M. and Foulard C. (1978) "Learning systems: Decision, Simulation and Control", Springer-Verlag, New york.

Favier G. et Di Martino M. (1984) "Commande adaptative d'untélépointeur",Colloque Commande Adaptative Aspects Pratiques et Theoriques, Grenoble 21-22 et 23 Novembre.

Favier G. (1987) "Computationally efficient adaptive identification algorithms, IEEE Int. Conf. on ASSP, Dallas, U.S.A.

FortescueT.R., Kershenbaum L.S. and Ydstie B.E. (1981) "Implementation of self-tuning regulators with variable forgetting factors", Automatica, vol. 17, n° 6, p.831.

Franklin J.N. (1968) "Matrix theory", Prentice-Hall, New Jersey.

Golub G.H. and Van Loan C.F. (1983) "Matrix computations", North Oxford Academic publishing Co. Ltd., Oxford.

Franklin G.F. and Powell J.D. (1980) "Digital control of dynamic systems", Addison-Wesley, London.

Goodwin G.C. and Sin K.S. (1984) "Adaptive filtering prediction and control", Prentice-Hall, Inc., New Jersy .

Hägglund T. (1983) "New estimation techniques for adaptive control", Departement of Automatic Contr., Lund Institute of Technol., Report LUTFD2/(TFRT-1025)/1-120/.

Ioannou P.A. and Kokotovic P.V. (1982) "Adaptive systems with reduced order models", lecture notes in control and information sciences, n° 47, Berlin, Springer Verlag.

Irving E. (1979) "Improving power network stability and unit stress with adaptive generator control", Automatica, vol. 24, p.1351.

Irving E. (1985) "Private communication".

Irving E., Falinover C.M. and Fonte C. (1986) "Adaptive generalized predictive control with multiple reference model", 2nd IFAC Workshop on Adaptive Systems in Control and Signal Processing, Lund Sweden.

Johnstone R.M. and Anderson B.D.O. (1982) "Exponential convergence of recursive least squares with exponential forgetting factor - adaptive control", Systems & Control Letters, vol. 2, n° 2, p.69.

Johnstone R.M., Johnson Jr.C.R., Bitmead R.R. and Anderson B.D.O. (1982) " Exponential convergence of recursive least squares with exponential forgetting factor", Systems & Control Letters, vol. 2, n° 2, p.77.

Kershenbaum and Fortescue T.R. (1981) "Implementation of on-line control in chemical process plants", Automatica, vol. 17, n° 6, p.777.

Koutchoukali M.S., Laguerie C. and Najim K. (1986) "Model reference adaptive control system of a catalytic fluidized bed reactor", Automatica. vol. 22, n° ·1, p. 101.

Kucera V. (1979) "Discrete linear control The polynomial equation approach", John Wiley & sons, New York.

Lakshmivarahan S. (1981) "Learning algorithms theory and applications", Springer-Verlag, New york.

Lam K.P. (1982) "Design of stochastic discrete time linear optimal regulators", Int. J. Systems, vol. 13, n° 9, p.979.

Landau I.D. (1979) "Adaptive control - The model reference approach", Marcel Dekker, New York.

Landau I.D. and Lozano R. (1981) "Unification of discrete time explicit model reference adaptive control designs", Automatica, vol. 12, n° 4, July, p.593.

Lee K.S. and Won-Kyoo L. (1985) "On-line optimizing control of a nonadiabatic fixed bed reactor", AIChE Journal, vol. 31, n° 4, p.667.

Ljung L. and Söderström T. (1983) "Theory and practice of recursive identification", MIT Press Massachusetts.

Maurath P.R. (1985) "Predictive controller design with applications to two-point composition control of distillation columns", Ph.D. Thesis, University of California, Santa Barbara, October.

Mendel J.M. (1973) "Discrete techniques of parameter estimation: the equation error formulation", Marcel Dekker, New York.

M'Saad M. (1987) "Sur l'applicabilité de la commande adaptative", Thèse d'Etat, I.N.P.G. Grenoble France.

M'Saad M., Duque M. and Landau I.D. (1986) " Practical implications of recent results in robustness of adaptive control schemes", 25th CDC, Athens, Greece, December .

M'Saad M., Duque M. and Irving E. (1987) "Thermal process robust adaptive control : An experimental evaluation", 10th I.F.A.C. World congress, München, F.R.G., July.

Najim K., Najim M., Koehret B. and Ouazzani T. (1976a) "Modelisation and simulation of a phosphate drying furnace", 7th annual Pittsburg conference on modeling and simulation, 26-28 April, Pittsburg, U.S.A.

Najim K., Najim M. and Zyoute M. (1976b) "Identification of a distributed parameter system: a phosphate drying furnace", 4th I.F.A.C. Symposium on identification and parameter estimation, 21-27 September, Tbilissi U.R.S.S.

Najim K. (1982) "Commande adaptative des procéssus industriels", Masson, Paris.

Najim K. et Muratet G. (1983) "Pratique de la régulation numérique des processus industriels", Masson, Paris.

Najim K., Dahhou B.,Youlal H. and Unbehauen H. (1985) "Adaptive control in chemical industry", IFAC/IFORS Conf. On Control Science and Technology for Development, Beijing, China, August 20-22.

Najim K., AL Khani S., Le Lann M.V. and Casamatta G. (1986a) "Self-tuning control of a pulsed liquid-liquid extraction column", I.F.A.C. Symposium on components, instruments and techniques for low coast automation and applications, 27-29 Nov., Valence, Spain.

Najim K., Le Lann M.V. and Casamatta G. (1986b) "Generalized predictive control of a pulsed liquid-liquid extraction column", Chem. Eng. Com., vol. 48, p. 237.

Najim K., Djaroud M., Le Lann M.V., Casamatta G. and Irving E. (1987a) "Long range predictive control with pole placement and double serie-parallel model reference of a Khüni column", 10th I.F.A.C. World congress, 27-31 July, München, F.R.G.

Najim K. , Le Lann M.V. and Casamatta G. (1987b) "Learning control of a pulsed liquid-liquid extraction column ",Chemical Engineering Science, vol. 42, n°. 7, p. 1619.

Najim K. , Le Lann M.V. and Casamatta G. (1987c) "Multivariable learning control of a pulsed liquid-liquid extraction column", 9th Int. Congress of Chemical Engineering, Chemical Equipment Design and Automation, CHISA'87, Praha, Czechoslovakia, August 30- September 4.

Najim K. et Muratet G. (1987) "Optimisation et commande en génie des procédés", Masson Editeur, Paris.

Najim K., Youlal H., Irving E. and Najim M. (1988) "Linear quadratic self-tuning control of a liquid-liquid extraction column", Optimal Control Applications and Methods.

Najim K. and Al Khani A. (1988) "Multivariable self-tuning control of a liquid-liquid extraction column", Seventh IASTED Int. Conf. on Modelling, Identification and Control, February 16-18, Grindelwald, Switzerland.

Narendra K.S. and Thathachar M.A.L. (1974) "Learning automata - A survey", IEEE Transactions on Systems, Man, and Cybernetics, vol. SMC-4, n° 4, p.323.

Naslin P. (1963) "Polynômes normaux et critère algébrique d'amortissement (I & II)", Automatisme, tome VIII, n° 6, Juin, p.215; n° 7-8, Juillet-Août, p.255.

Praly L. (1983) "Robustness of indirect adaptive control based on pole placement design", 1th IFAC Workshop on adaptive systems in control and signal processing, San Francisco,USA

Praly L. (1984) "Robustesse des algorithmes de commande adaptative", Colloque commande adaptative: Aspects pratiques et théoriques, Novembre, Grenoble.

Praly L., Najim K., Youlal and Najim M. (1988) "Direct adaptive control of a pulsed liquid-liquid extraction column", Int. J. Adaptive Control and Signal Processing.

Poznyak A.S. (1975) "Investigation of the convergence of algorithms for the functioning of learning automata", Automation and remote control, vol. 36, n° 1, part 1, January, p.77.

Robbins H. and Monro S. (1951) "A stochastic approximation method", Ann. Math. Stat., vol. 22, P. 400.

Samson C. (1983) "Stability analysis ofadaptively controlled systems subject to bounded disturbances", Automatica, vol. 19, n° 1, p.81.

Seborg D.E., Edgar T.F. and Shah S.L. (1986) "Adaptive control strategies for process control: a survey", AIChE Journal, vol. 32, n°6, June, p.881.

Seret D. et Macchi O. (1982) "Automates adaptatifs optimaux", Technique et Science Informatique vol. 1, n° 2, p.143,

Stoica P., Eykhoff P., Janssen P. and Söderstöm T. (1986) "Model structure selection by cross-validation", Int. J. Control, vol. 43, n° 6, p.1841.

Thathachar M.A.L. and Ramakrishnan K.R. (1981) "A hierarchical system of learning automata", IEEE Trans. on Systems Man, and Cybernetics. vol. SMC-11, n° 3, p. 236, March.

Thathachar M.A.L. and Harita B.R. (1987) "Learning automata with changing number of actions", IEEE Trans. on Systems Man, and Cybernetics. vol. SMC-17, n° 6, p. 1095, December.

Tuffs P.S. and Clarke D.W. (1985) "Self-tuning control of offset: a unified approach", I.E.E. Proc. vol. 132, part D, n° 3, p.100.

Wellstead P.E. and Sanoff S.P. (1981) "Extended self-tuning algorithm", Int. J. Control, vol. 34, p.433.

Wittenmark B. (1979) "Self-tuning PID-controllers based on pole placement", Departement of Automatic Contr., Lund Institute of Technol., Report LUTFD2/(TFRT-7179)/1-037/.

Ydstie B.E. and Sargent R.W..H. (1982) "Convergence and stability properties of an adaptive regulator with variable forgetting factor", IFAC Symposium On Identification and System Parameter Estimation, Washington DC, June 7-11.

Ydstie B.E. (1982) "Robust adaptive control of chemical processes", Ph.D. Thesis, University of London.

Youlal H., Najim K. and Najim M. (1988) "Regularized pole placement adaptive control", IFAC Workshop on Robust Adaptive Control, 22-24 August, Australia.

Ziegler J.G. and Nichols N.B. (1943) "Optimum settings for automatic controllers", Trans. ASME, vol. 65, p.433.

INDEX